MIRROR-IMAGE
ASYMMETRY

MIRROR-IMAGE ASYMMETRY

AN INTRODUCTION TO THE ORIGIN AND CONSEQUENCES OF CHIRALITY

JAMES P. RIEHL

A JOHN WILEY & SONS, INC., PUBLICATION

Published by John Wiley & Sons, Inc., Hoboken, New Jersey.
Published simultaneously in Canada.

For general information on our other products and services or for technical support,
please contact our Customer Care Department within the United States at (800)
762–2974, outside the United States at (317) 572–3993 or fax (317) 572–4002.

Wiley also publishes its books in a variety of electronic formats. Some content that
appears in print may not be available in electronic formats. For more information about
Wiley products, visit our web site at www.wiley.com.

Library of Congress Cataloging-in-Publication Data:

Riehl, James P.
    Mirror-image asymmetry : an introduction to the origin and consequences
of chirality / by James P. Riehl.
        p. cm.
    Includes bibliographical references and index.
    ISBN 978-0-470-38759-7 (pbk.)
    1. Geometry in nature. 2. Symmetry (Biology) 3. Chirality. 4. Random dynamical
systems. I. Title.
    Q172.5.S95R54 2010
    500–dc22

                                                                2009051005

Printed in the United States of America

10 9 8 7 6 5 4 3 2 1

To Cecelia, Patrick, and James;
and in memory of my mother Alice

CONTENTS

PREFACE

It is perhaps easy for some to not notice the symmetry or asymmetry in their everyday lives, but for someone like me, who has spent almost 35 years developing and using chemical and physical methods to study asymmetric molecules, to not see this aspect in my everyday view of the macroscopic world is impossible. There is certainly beauty in symmetry, whether it is in the spherical, radial, or bilateral symmetry of physical objects, or in the mathematics and equations that one needs to describe the most fundamental laws of nature. However, if one is able to understand the purposeful or programmed departure from symmetry, the world in my view becomes much more interesting and no less beautiful.

This book has been written in a way that parallels the development of my understanding and interests in chirality. It begins with the necessary definitions and language that is needed to communicate the structure of molecules related through their mirror image, and then, in Chapter 2, describes the early history of chirality in molecules found in living systems. Chapter 3 is a short summary of the various theories of the origin of chirality, and in Chapters 4 and 5 we present information concerning the role that chiral molecules play in the pharmaceutical industry and in sensory recognition and biochemical control. Beginning in Chapter 6 we leave the molecular consequences of chirality, and begin a description of chirality in large-scale living systems. Chapter 7 deals with the right-handedness of human beings, and Chapter 8 is concerned with the consequences of a dominantly right-handed population in sports, windmills, and many other aspects of the chiral world that we inhabit.

The reader unfamiliar with basic chemistry and the way that chemists draw molecules should not be waylaid by all of the chemical drawings that are given in early chapters. Certainly, the ideas and principles may be understood without appreciation for the three-dimensional structures. The reader is especially encouraged to continue forward to Chapters 6–8, where photographs and diagrams of familiar objects are used to illustrate how chirality is present in the macroscopic world. Many readers who are far removed from their chemistry studies may benefit from the quick review of chemical structural drawings given in the short appendix at the end of the book.

ACKNOWLEDGMENTS

Since this book has evolved from a long period of interest in chirality, many former mentors, colleagues, and students have contributed to my ability to begin and ultimately finish this project. Fred Richardson at the University of Virginia introduced me to this area of research as a young postdoctoral fellow, and colleagues Harry Dekkers at the University of Leiden, and Janina Legendziewicz at the University of Wrocław were most important in helping this theoretician latch on to real experimental problems. Thanks to all of my graduate students, postdoctoral fellows, faculty colleagues and collaborators, and especially to my former doctoral students Gary Hilmes and Christine Maupin, who have become essential friends and colleagues. In Duluth, continuing research progress would

have been impossible without the efforts of my recent postdoc
Gilles Muller and his wife Françoise Muller.

Much of this book was written while I was on an administrative leave in Paris in the Fall 2007, thanks to the support of
Chancellor Katherine Martin and Vice Chancellor Vince Mag-
nuson of the University of Minnesota Duluth. This time away
from my position as Dean of the Swenson College of Science and
Engineering at UMD was only possible because of the great
work and dedication of Tim Holst, who served as Dean in my
absence. A dedicated staff in the Dean's Office at UMD (Stan,
Janny, Tricia, Sue, Connie, Lorraine, Mae, Amy, Sally, Brenda,
and Lori) helped immensely in giving me the time to complete
the writing of this manuscript. I also need to express
my appreciation to Rachel MaKarrall, Brett Groehler, and
Deb Shubat for their assistance in the preparation of illustra-
tions and photographs, and to Dr. Robert Carlson (UMD) and
Dr. Marta Laskowski (Oberlin) for reviewing parts of this
manuscript. My editor, Anita Lekhwani, has been a source of
constant support and encouragement throughout the process
of getting this project to completion.

Special thanks to my wife, Cecelia, for her personal support,
and for her efforts in reviewing the manuscript.

LIST OF FIGURES AND BIOGRAPHIC PHOTOGRAPHS

FIGURES

BIOGRAPHIC PHOTOGRAPHS

THE MIRROR IMAGE

In at least one version of the story of the Greek mythological hero Narcissus, he becomes infatuated with his own reflection in a pool of water, and does not recognize that this is just himself until he tries to kiss his image in the water. Narcissus becomes so distraught at this discovery that he kills himself. Fortunately, humans do not seem to have any difficulty recognizing that our reflection in a mirror or pool of water is not a friend or an enemy. The fact that in our "mirror image" our left hand becomes our right hand, our hair is parted on the other side, and the logo on our shirt has "backward" letters (STATEU becomes UƎTATƨ) is normal once we get past the fascination with mirrors of our early childhood. Our image of ourselves is really our mirror image rather than our true image (the one that others see) because this is the one that we see every day in the mirror.

As scientists have become more capable of probing the structure of three-dimensional objects at the molecular and atomic levels, the need to understand the concept and consequences of mirror-image symmetry has increased enormously.

Mirror-Image Asymmetry: An Introduction to the Origin and Consequences of Chirality
by James P. Riehl
Copyright © 2010 John Wiley & Sons, Inc.

As we will see, there are very important aspects of nuclear physics that directly involve mirror-image relationships between fundamental particles, and to a chemist, knowledge of the mirror-image symmetry of molecular structures has become essential in efforts to understand the mechanisms of biological processes and drug activity. The presence or lack of mirror-image symmetry has also become an important probe in developmental biology and related areas. The macroscopic world in which we live also has many important and interesting properties that are connected to mirror-image symmetry. We begin this book and this chapter with a number of definitions, and some nomenclature.

HANDEDNESS

Throughout this book you will see the terms *right-handed* and *left-handed*. What do we mean when something is called *right-handed*? For a human being, the answer is pretty clear. It means that people use their *right hand* predominately for actions that they perform daily. But we all know people who may throw a ball with their right hand, and perhaps play golf the way most right-handers do, but write with their left hand. There are certainly degrees of handedness among people, and the question sometimes becomes not whether a person is right- or left-handed, but the extent or degree of the person's right- or left-handedness. If you throw a ball, raise your hand, comb your hair, sign your name, brush your teeth, and point with your right hand, then you are said to be right-handed. You probably also kick with your right foot, fold your arms with your right arm on top, and wink with your right eye. As we see in Chapter 7, most of us *Homo sapiens* are dominantly right-handed, and we call ourselves right-handed, but there are obviously people who are very left-handed, and a range of people in between.

Of course, the words *left* and *right* have interesting connotations. *Right* (*rectus* in Latin) is associated with strong, correct,
clever, and so on, and *left* (*sinister* in Latin, *gauche* in French)
connotes evil, clumsy, untactful, etc. In this discussion we are
omitting all the possible political interpretations of these two
terms (see Sidebar 1.A). In this book we are interested in
studying the absolute nature of right- and left-handed symmetry in our world, and the first thing we should recognize is that
we should limit as much as possible the use of the terms *left-
handed* and *right-handed* unless we have very clear definitions of
what these terms mean. In fact, maybe we should only apply
these terms to people or other mammals that have hands!

Sidebar 1.A

> **Right-Wing ↔ Left-Wing Politics.** In politics, the terms *left
> wing* and *right wing* originated from the seating arrangement
> in various legislative bodies in France during the French
> revolutionary period at the end of the eighteenth century.
> The aristocracy sat to the right of the leader or speaker, and
> the commoners sat on the left. At this time in history, right-
> wing politics were associated with the interests of the church
> and royalty; the left wing was opposed to these interests, and
> represented the rising capitalist class, the so-called bour-
> geoisie. So the left wing was supporting laissez-faire capi-
> talism and free markets. This is generally opposite to what
> would be called "left wing" in modern times. (Wikipedia.)

So, how do we study and talk about this important property
of our world in an unambiguous way? We first need a term to
apply to a static or dynamic system in which this type of mirror-
image symmetry is present. We could and will use the term
handed, but for the reasons given above, we will more often use
the slightly more general term *chiral* (pronounced *kai*-ral)

Left hand   Right hand

FIGURE 1.1. A left hand and a right hand as mirror images.

and call this property *chirality*. Not surprisingly, this word is derived from the Greek word χειρ (cheir), meaning *hand*. The first use of this term and its definition occurred in a lecture by Lord Kelvin (Sir William Thomson) in 1884 [1]. Very specifically, a system is *chiral* if the mirror image of the system is not "superimposable" on the original system. The best example is, of course, the human hand which we depict in Figure 1.1. The mirror image of the right human hand is the left human hand. These are distinct structures and are not identical. In this very special case, we can certainly talk about a right hand and a left hand, and know exactly what we mean.

In Figure 1.2 we show the picture of a house, and its mirror image. These two structures are obviously not identical, but we really can't describe one of them as a right-handed house, and the other as left-handed. We can describe the house by saying that the door or window is on the left or right as we face the front, and so on, but in this case, of course, we don't really need a nomenclature. We will call knowing exactly which mirror

FIGURE 1.2. Mirror-image houses.

image we are referring to as "knowing the *absolute* structure." In fact, in some situations for molecules it is not really necessary, or in other situations for various reasons it is not even possible to know the absolute structure. At the microscopic molecular level, however, a nomenclature that allows one to describe an absolute three-dimensional (3D) structure is essential, if we are attempting to communicate structural information in order, for example, to understand how biologically important chiral molecules interact and function. We will discuss the importance of knowing the absolute structure in Chapter 2, but for now we will focus on introducing the basis for the classification of 3D structures as chiral, and discuss the nomenclature that chemists and biologists use to identify which mirror-image molecule they are talking about. The various nomenclature systems will be presented here in some detail, but some readers might choose simply to scan this section to get some idea of the basis for this aspect of chiral structure.

THE SYMMETRY OF NONSUPERIMPOSABLE MIRROR IMAGES

With just a little practice it is easy in most cases to determine whether an object has a nonsuperimposable mirror image—in other words, whether the object is chiral. This is especially true for objects that have "no" symmetry, or more properly no *symmetry elements*. The concept of symmetry elements is an important one for many scientists and mathematicians. In the language of symmetry theory an object is said to possess a *set of symmetry elements*. For our purposes we will focus primarily on the existence of symmetry planes and inversion centers, and refer the reader to the many textbooks dealing with the concept of symmetry groups for a more complete analysis of that topic. Some suggested books on symmetry classifications and theory are listed at the end of this chapter.

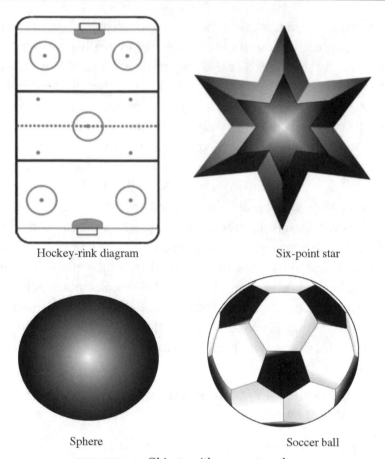

Hockey-rink diagram   Six-point star

Sphere   Soccer ball

FIGURE 1.3. Objects with symmetry planes.

A three-dimensional object possesses a symmetry plane, if reflection of the object through the plane produces the other half of the object. Obviously, the plane must pass through the center of the object. Examples of objects that possess symmetry planes are shown in Figure 1.3. This might not be so obvious, but every object that has a plane of symmetry is *achiral*, that is, *not* chiral, and the mirror image is superimposable. You can see this with the objects depicted in Figure 1.3. You will have to imagine taking the mirror image and rotating it around (if necessary) to see that it is identical to the original object. If an object possesses an

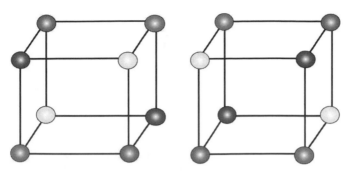

FIGURE 1.4. Mirror-image cubes with inversion symmetry. See color insert.

inversion center then it also will *not* be chiral. An example of this special case is given in Figure 1.4, where we show a cubic structure with colored balls at the corners. The center of inversion is in the middle of the cube. For every part of this structure, for example, a red ball, an identical part (another red ball) is located on a line through the center and out the other side at the exact distance from the center. We give both the structure and its mirror image, and you should be able to convince yourself that these two structures are superimposable, and therefore not chiral. There are also some very special cases of objects that have no symmetry planes, and no inversion center, but are still achiral. These extraordinarily rare structures possess a rotation/reflection symmetry element, but this topic will not be discussed here [2].

A major focus of this book is the connection between the microscopic molecular world and the macroscopic world that we live in. With this in mind, we need to describe the nomenclature used for molecular structures that are chiral. It is sometimes useful (but not universally applied) to classify chiral structures that have *no* symmetry elements as *asymmetric* and those that have some symmetry elements but are still chiral as *dissymmetric*.[1] Using this distinction,

---

[1]Formally, in the theory of symmetry groups, every structure has one trivial element, called the *identity element*, which simply does nothing to the structure.

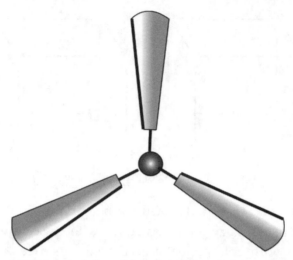

FIGURE 1.5. A three-bladed right-handed propeller.

we would classify the hand and house pictured in Figure 1.2 as being asymmetric. An example of a dissymmetric structure is the propeller shown in Figure 1.5. You can see that the propeller has no planes of symmetry or an inversion center, but what it does have is a *rotation axis*. Spinning this three-bladed propeller by 120° around an axis oriented perpendicular to the plane of the page yields the identical structure. One way to understand this is to look at an object and then, closing your eyes, have someone perform the symmetry operation. If after opening your eyes you can't tell whether the symmetry operation has been performed, then this is, indeed, a symmetry element. In the specific case of the three-bladed propeller, we say that the object has a threefold rotation axis, since it would take three rotations of 120° to get back to the original structure. Starting with the Wright brothers' plane, which used two two-bladed propellers, the absolute structure of the propellers used is an important consideration (see Sidebar 1.B).

Sidebar 1.B

Wright Propellers. The Wright brothers spent several years designing and testing efficient propellers after realizing that the existing propellers used for watercraft would not work in their new flying machine. This photograph below, which was taken at the Air and Space Museum in Washington, DC, shows that the two propellers that pushed, rather than pulled, the machine were mirror images of each other. This design serves to compensate for a force that would tend to cause the airplane to twist, and thus makes their airplane net achiral. In modern airplanes with two or more propellers, the propellers all turn in the same direction. This is convenient, of course, since the airplane mechanic doesn't need to worry about the side of the plane on which the engine is being installed. Modern airplane control systems are much more sophisticated than those of the Wright brothers, and it is easy to compensate for any net twisting force because all the propellers are turning in the same direction.

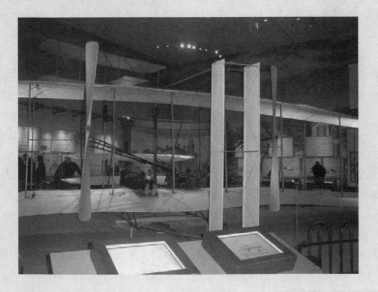

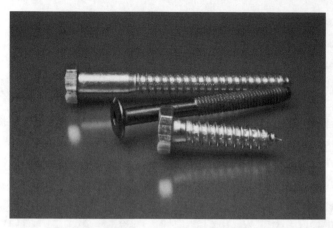

FIGURE 1.6. Right-handed screws. (Photographed by Brett Groehler.)

The nomenclature used to describe the absolute structure of a dissymmetric structure such as a propeller is somewhat easy to understand, because we have experience with the terms used to describe the threads on screws and pipes. A *right-handed* screw is defined as one that if turned clockwise penetrates the substance being screwed into. [We will discuss the terms *clockwise* and *counterclockwise* (or *anticlockwise*) in Chapter 7.] Figure 1.6 shows a few right-handed screws. The common definition of a *right-handed propeller* is one in which the blade, as you go from front to back, tilts in a clockwise direction. Propellers are, of course, normally driven by a motor, and the definition that you will often find for a *right-handed* propeller is one that, when viewed from the back, if driven clockwise will push the air or water out toward the front of the propeller, which, of course, is pointed out the back of the vehicle. If you think this is confusing...well, you're correct! The propeller shown in Figure 1.5 is right-handed.

Why is the term *right-handed* used for screws and why are most screws *right-handed*? It's because most of us are right-handed, and if you are trying to drive a screw into something with some force, your hands, if you are right-handed, do a much better job when you turn the screw clockwise. So we call this natural direction or screw chirality *right-handed*. This seems to be a pretty loose connection between the fact that most people are right-handed, and the direction that a screw turns, but it is the basis for this nomenclature. It is useful sometimes to think about how one would describe this to an interested alien over a voice-only line! You will see that the use of the term *right-handed* or *left-handed* for all the cases described below may give the wrong impression that the various types of chirality that are sometimes considered as right-handed have some fundamental symmetry property in common.

Rather than calling this form of mirror-image symmetry right-handed or left-handed, some other notations have been defined. In various contexts for this type of dissymmetry, scientists use P or M (which actually stand for plus and minus), the uppercase Greek letters lambda (Λ) or delta (Δ), and *R* or *S* (rectus or sinister). Our intention here is not to concentrate on the precise naming of molecules, but it is important to recognize the utility, proper use, and limitations of this nomenclature. We begin by noting that these nomenclatures all deal with a three-dimensional *structure* and not some measurable property, as we will see below.

NOMENCLATURE FOR CHIRAL MOLECULES

A large number of important chemical compounds contain a central carbon atom (C) and four attached substituents oriented to the alternating corners of a cube as shown in Figure 1.7. The overall geometry is that of a tetrahedron, and the carbon atom is referred to as being *tetrahedral*. Chemists have worked hard at defining an unambiguous nomenclature for chiral carbon atoms since it was discovered by the Dutch graduate student Jacobus Henricus van't Hoff (Biographic Photo 1.1) and independently by the French chemist Joseph-Achille Le Bel (Biographic

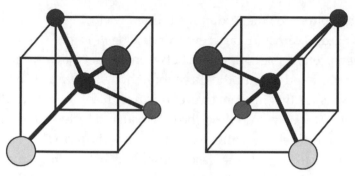

FIGURE 1.7. Tetrahedral enantiomers. See color insert.

Photo 1.2) in 1874 that carbon atoms could be tetrahedral and that this led to mirror-image isomers (see Sidebar 1.C). Most of the nomenclature associated with the naming of chiral molecules actually preceded knowledge of the absolute structure. In fact, the first determination of the absolute structure of a chiral molecule containing an asymmetric carbon didn't occur until 1951!

BIOGRAPHIC PHOTO 1.1. Jacobus Henricus van't Hoff. (Permission for use granted from the Edgar Fahs Smith Collection at the University of Pennsylvania Library.)

BIOGRAPHIC PHOTO 1.2. Joseph-Achille Le Bel. (Permission for use granted from the Edgar Fahs Smith Collection at the University of Pennsylvania Library.)

Sidebar 1.C

Jacobus Hendricus van't Hoff. While a graduate student at the University of Utrecht, van't Hoff was trying to understand experimental results on how many compounds existed with the same chemical formula (isomers). For example, there appeared to be only one molecule with the formula CH_2Cl_2 (i.e., one carbon, two hydrogen, and two chlorine atoms). His explanation was that four substituents on carbon were arranged in a tetrahedron. He also concluded that, if correct, this would lead to mirror-image isomers for molecules with four different substituents. This idea was at first ridiculed by his senior colleagues, but, now, of course, is recognized as the basis for organic (carbon-based) molecular structure. He was the recipient of the first Nobel Prize in Chemistry for this contribution in 1901.

Sidebar 1.D

Origin of the Cahn–Ingold–Prelog Rules. After a scientific meeting organized by the Royal Chemical Society in Manchester, England in 1954, a dance was organized. While most participants were dancing, a few drank beer and discussed chemistry. Among them were the Royal Society President Sir Christopher Ingold, the editor of the *Chemical Society Journal*, Robert Cahn, and Swiss chemist Vladimir Prelog. Apparently, after a "vigorous" criticism by Dr. Prelog of a recent paper, he was invited by Drs. Ingold and Cahn to join them in sorting out this nomenclature. After several meetings, the principles of the new nomenclature system were published in the Swiss journal *Experientia* in 1956. (Current Contents, Dec. 13, 1982, p. 18.)

In order to name these structures in an unambiguous way, European chemists Robert S. Cahn, Sir Christopher Ingold, and Nobel Prize winner Vladimir Prelog (see Sidebar 1.D) devised a system (often called the CIP system) in which the chirality of a chiral carbon atom is based on an ordering of the four substituents using established priority rules depending on atomic number (the highest atomic number equals the highest priority, i.e., 1). Additional rules are used to establish priority numbers for situations in which the connected atoms are the same, but differ in their attachments. Once the priority numbers are determined, the structure is oriented such that the lowest priority number (4) is aligned to the rear of the structure, and determining whether the priority order of the other three substituents represents a clockwise (*R*) or counterclockwise (*S*)

FIGURE 1.8. Bromochlorofluoromethane enantiomers.

direction in going from 1 to 2 to 3. This will make sense only with an example! Let's consider a real chiral molecule with the formula CHFClBr as shown in the two mirror-image structures drawn in Figure 1.8. From the periodic table of the elements we know that the atomic numbers of these elements are $H = 1$, $F = 9$, $Cl = 17$, and $Br = 35$. With this knowledge we order the substituents in Figure 1.8 and orient the molecules so that H, the lightest and therefore lowest-priority molecule, is oriented away from us. This is done in Figure 1.9. Hopefully, in the molecule on the left, you can see how proceeding from 1 to 2 to 3 involves a clockwise motion, and similarly on the right, a counterclockwise rotation.

In most cases the R,S notation is straightforward and un-ambiguous, but it is important to realize that the connection between notation and structure is often ill-defined. In other words, the symbol R or S is not always useful for comparison of similar structures. For example, if chlorine (Cl) is replaced by iodine (I, atomic number = 53), then the structures, priority, and names of these compounds would be as shown in Figure 1.10. So the structures on the left (or right) in Figures 1.9 and 1.10 are very similar, but in the accepted nomenclature have different designations for the absolute chiral structure. If a researcher is interested in reaching conclusions based on

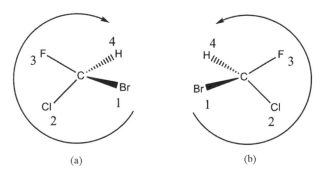

FIGURE 1.9. Tetrahedral chiral nomenclature: (a) (R)-bromochlorofluoro-methane; (b) (S)-bromochlorofluoromethane.

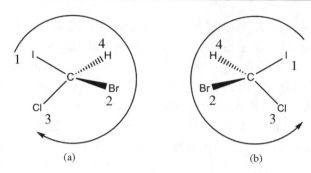

FIGURE 1.10. Enantiomers of iodobromofluoromethane: (a) *S*-iodobromo-fluoromethane; (b) *R*-iodobromofluorormethane.

comparative structures, then, although the CIP nomenclature is clear and exact, it may lead to some confusion if nomenclature rules are not completely understood and applied correctly.

Preceding the formulation and introduction of the CIP nomenclature, and decades before absolute structures were known, Professor Emil Fischer of the University of Würzburg (later the University of Berlin) suggested a system in which chiral compounds were related to chemical transformations of the known compound glyceraldehyde. At this time, two forms of glyceraldehyde were known. These two forms were identical in composition and had identical physical properties. The only difference was that one form (isomer) rotated the plane of a polarized beam of light to the left, and the other isomer rotated the plane of polarization to the right. We will discuss the measurement and causes of optical rotation in Chapter 2, but, very briefly, if a plane-polarized beam of light passes through a container containing a chiral substance, then the plane of the polarization will be rotated to either the left or the right as shown in Figure 1.11. In this figure the plane of polarization, indicated by the arrow, is rotated in a clockwise direction (viewed from the source looking in the direction in which the light is traveling) as the beam passes through the sample containing chiral molecules. A rotation in this direction would

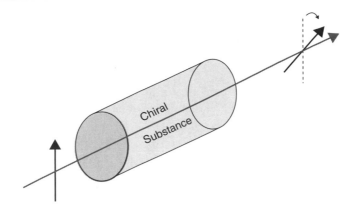

FIGURE 1.11. Schematic diagram of optical rotation. See color insert.

be designated as positive, and the molecule under study would have the prefix (+) attached to the name. The fact that certain molecules or mixtures could rotate the plane of polarized light led to the separation of substances into those that were called "optically active" and those that were called "optically inactive." These terms are still used today to describe whether a molecule is chiral (optically active).

So the two forms of glyceraldehyde that Fischer studied were called (+)-glyceraldehyde and (−)-glyceraldehyde. A convention was established for orienting the molecule to be named. Fischer decided on the nomenclature presented in Figure 1.12. The conventional drawings given in this figure are intended to represent the tetrahedral geometry of a carbon atom

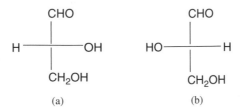

(a)   (b)

FIGURE 1.12. Fischer projections of glyceraldehyde enantiomers: (a) D-(+)-glyceraldehyde; (b) L-(−)-glyceraldehyde.

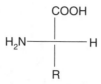

FIGURE 1.13. Structure of L-amino acids.

with the solid horizontal line in front of the plane of the paper and the vertical line representing substituents in the back of the plane of the paper. The enantiomeric identity denoted by small capital letters is D and L. In 1891, how did Fischer know that the structure of glyceraldehyde that gave a positive rotation was the D form shown? He actually couldn't have known this until the absolute structure of a chemically related compound was determined by X-ray crystallographic methods in 1951. He had a 50% chance of getting it right, and it turns out that he did!

There are problems with this older nomenclature that led to the CIP system, but this D,L notation is based on structures and not atomic numbers, so it is still used in some applications. Of special interest is the structure of chiral amino acids. The proteins of the human body are composed of long polymeric chains containing a sequence consisting of only 20 amino acids. Amino acids have the general structure given in Figure 1.13 (using the Fischer "projection" method of drawing them). The R in this figure denotes 1 of the 20 different molecular fragments or *residues*, and not the R as in R/S. (This might be a bit confusing!) When R is H (hydrogen), this is the simplest amino acid, glycine (see Figure 1.14) and is not chiral, because two of the substituents are the same, two hydrogens. In the nomenclature introduced by Fischer the other 19 amino acids are all L-amino acids. That's right; all the chiral amino acids in our proteins are L! Do you think that they all rotate light to the left? In other words, are they all L-(−)? The answer is "no."; 8 are L-(+), 11 are L-(−), and, of course, glycine, where the R group is H does not rotate the plane

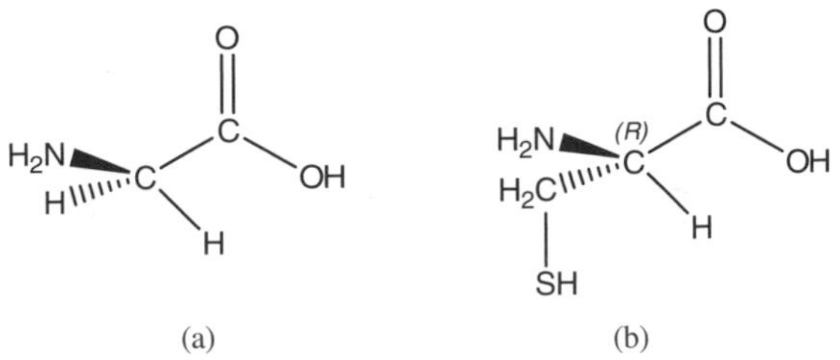

FIGURE 1.14. Molecular drawings of (a) glycine and (b) L-cysteine.

at all since it is not chiral. How about the R,S nomenclature? In fact, in all of the L-amino acids the central carbon atom is an S structure except cysteine (see Figure 1.14) because in this amino acid, one of the carbon atoms attached to the chiral tetrahedral carbon is attached to a sulfur atom, S (atomic number 16). This changes the priority numbering, the order is changed, and this amino acid is denoted R-cysteine. Chemists often used the term *stereochemistry* when they are studying the three-dimensional structure of chiral compounds.

NOMENCLATURE FOR DISSYMMETRIC AND OTHER CHIRAL MOLECULES

Although tetrahedral carbon is probably the most important and certainly the most common chiral structure, many other chiral molecular structures can be classified as asymmetric or dissymmetric. In the very late nineteenth century, the "father of coordination chemistry" Professor Alfred Werner of the University of Zurich developed a theory for the structure of compounds where a central atom other than carbon could form bonds to different numbers of atoms. This "coordination theory" was most successful and applicable to structures in which the number of bonds (the *coordination number*) was 6. Werner was very familiar with the work of van't Hoff and Le Bel on tetrahedral carbon, and he spent many years trying to synthesize and separate compounds that were based on octahedral geometry.

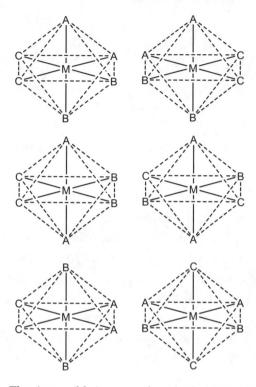

FIGURE 1.15. The six possible isomers of octahedral $MA_2B_2C_2$. The top two isomers are chiral.

It was easy to see that in an octahedral structure four different compounds could be made with a central metal atom with two each of three different substituents. This compound would have the general formula $MA_2B_2C_2$. Different compounds that could be formed with identical formula are called *isomers*. It was also easy to see that two of the six possible isomers were chiral. All of the possible isomers of octahedral $MA_2B_2C_2$ are drawn in Figure 1.15. The solid lines in this figure represent the bonds from the central metal atom to the substituents, and the dashed lines have been added to show the outline of the octahedron, which contains eight triangular faces. A real proof of the coordination theory of Werner was the successful measurement of optical rotation in octahedral complexes in 1911 [3].

Modern rules for the naming of structures such as those given in Figure 1.15 are based on the CIP [R/S] system introduced above for carbon. It is still the case that the clockwise or anticlockwise (i.e., *counterclockwise*) sequence of bonded atoms on the basis of their priority number is used to define the overall chirality of the structure, but using the symbol C (clockwise) and A (anticlockwise) instead of R and S is now recommended. The rules for determining priority are unchanged. In octahedral structures, the reference axis is the one that contains the highest-priority atom, with the lowest-priority atom on the opposite side of the metal. One then looks from the highest-priority substituent toward the central metal atom, and determines the priority sequence in the perpendicular square plane. In these cases, the clockwise direction from high to low sequence is designated as C and the anticlockwise one as A. This is illustrated in Figure 1.16 for the octahedral complex [$Co(CN)_2(NH_3)_2(H_2O)_2$]$^+$. The priority numbers (1 2 3) are assigned according to the atomic numbers of oxygen (8), nitrogen (7), and carbon (6). The sense of rotation is shown around the axis determined from the line from H_2O to Co to CN.

The CIP nomenclature has been applied to virtually all chiral molecules, such that chemists are able to communicate to each other the absolute structure of such species in an unambiguous

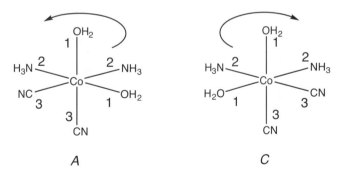

FIGURE 1.16. A and C enantiomers of [$Co(CN)_2(NH_3)_2(H_2O)_2$]$^+$.

way. The International Union of Pure and Applied Chemistry
(IUPAC) maintains a Website (www.iupac.org/publications/
books/seriestitles/nomenclature.html) containing links and
reference material on current nomenclature recommendations.
As we saw for amino acids as described above, however, some of
the older nomenclatures or conventions commonly used for
specific classes of chiral molecules are somewhat more useful,
since they are based on structural similarities.

As introduced above, dissymmetric structures include propellers, screws, and other "twisted" structures. In this category,
of particular importance to modern biochemistry and molecular biology is the helix, so we will spend some time describing
the chirality of helices as drawn in Figure 1.17.

The definition of chirality of helices is the same as that for
screws given above. More formally, a right-handed helix is one
that turns clockwise as you move along the length of the helix.
This is often related to a right hand. If you move your right
thumb along the long axis of a helix, then the curvature is the
same as that of your curled fingers. By definition, right-handed
helices are called P for plus [of course, the plus sign ($+$) means
something different!], and this is also designated as being Δ.
Left-handed helices are M and Λ. Most importantly, DNA,
which is actually a double helix, is normally right-handed.

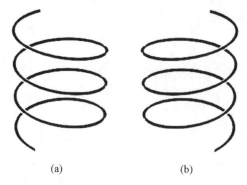

(a)   (b)

FIGURE 1.17. Nomenclature for helices: (a) left-handed helix, M or Λ; (b)
right-handed helix, P or Δ.

Sidebar 1.E

Left-Handed DNA. Although it is possible for DNA to occur in a left-handed helix (the so-called z form, which is not a mirror image of right-handed DNA), the overwhelmingly dominant natural form of DNA is right-handed. This is so important and so fundamental to molecular biology that it is amazing how many times DNA is drawn with the incorrect helicity. Dr. Thomas Schneider, a research biologist at the Center for Cancer Research Nanobiology Program at the National Institutes of Health, has long maintained a Website of errors in the depiction of DNA. His Website is named the Left-Handed DNA Hall of Fame (http://www.lecb.ncifcrf.gov/~toms/ LeftHanded.DNA.html). My favorite is the October 23, 1998 cover of *Science* magazine, the "Genome Issue": (Permission for use granted from *Science* magazine. See color insert.)

SUMMARY

Our world is full of chiral objects from the microscopic molecular to the macroscopic! This is a simple and obvious conclusion that everyone in our world can make. However, it takes more work and thought to define and communicate the exact (absolute) three-dimensional structure of objects in our world. In certain situations it suffices to know that something is chiral, but a complete understanding of biological and chemical processes requires that one be exact in the articulation of chiral structures. We have described several ways of doing this, and exactly how we choose to define a chiral structure varies a bit with the particular system under investigation and the nature of the asymmetry or dissymmetry. In virtually every case, our frame of reference for nomenclature is the fact that we are "handed," and objects are often referred to as being right-handed or left-handed. Hopefully, it is obvious to the reader of this chapter that the connection between right-handed people and right-handed helices or right-handed amino acids and so on is, however, not very meaningful.

SUGGESTIONS FOR FURTHER READING

Vincent, A., *Molecular Symmetry and Group Theory: A Programmed Introduction to Chemical Applications*, 2nd ed., Wiley, Chichester, UK, 2001.

Bishop, D. M., *Group Theory and Chemistry*, Dover Publications, Mineola, NY, 1993.

Carter, R. L., *Molecular Symmetry and Group Theory*, Wiley, Chichester, UK, 1997.

North, M., *Principles and Applications of Stereochemistry*, Stanley Thornes Ltd., 1998.

Heilbronner, E. and J. D. Dunitz, *Reflections on Symmetry*, VCH Publishing, Basel, 1993.

REFERENCES

1. Kelvin, W. T., The second Robert Boyer lecture, *J. Oxford Univ. Junior Sci. Club* **18**: 25 (1884).

2. Cotton, F. A., *Chemical Applications of Group Theory, 3rd ed.*, Wiley, New York, 1990.

3. Werner, A., V. L. King, and E. Scholze, Zur Kenntnis des asymmetrischen Kobaltoms. *I, Ber. Deutsche Chem. Gestalt.*, **44**: 1887–1898 (1911).

LIST OF BIOGRAPHIC PHOTOGRAPHS, SIDEBARS, AND FIGURES

MOLECULAR CHIRALITY IN LIVING SYSTEMS

In reading about the history of science, we often find that famous scientists in previous eras made important contributions in many different fields. Such is the case for the French scientist Jean Baptiste Biot (Biographic Photo 2.1), whose name is associated with major discoveries in mathematics, astronomy, physics, and chemistry. Biot discovered in 1815 that the phenomenon of rotation of the plane of polarized light (a topic introduced and described very briefly in Chapter 1), which was known for quartz crystals, could also be observed in naturally occurring liquids such as oils of lemon and turpentine [1]. We now know that these two substances are not pure compounds, but complex mixtures. The main constituent in lemon oil is (+)-limonene, sometimes referred to as *d-limonene* (a nomenclature that we will discuss shortly), and the main component in "French" turpentine is (+)-α-pinene. It is interesting to note that the molecule present in turpentine in North America is (−)-α-pinene. The structures of these compounds are given in Figure 2.1. Note that the molecule limonene has one asymmetric

Mirror-Image Asymmetry: An Introduction to the Origin and Consequences of Chirality by James P. Riehl
Copyright © 2010 John Wiley & Sons, Inc.

Biographic Photo 2.1. Jean Baptiste Biot.

carbon center, and α-pinene has two chiral centers.[1] Biot's successful but problematic measurement of the same phenomenon with gases is the subject of a legendary experiment gone bad (see Sidebar 2.A). Biot and his scientific contemporaries soon realized that almost all substances that were "organic" (i.e., derived from living systems) rotated the plane of polarized light.

[1]For the sake of correctness and completeness, we will give chemical structures for a number of important compounds in this chapter, but we realize that not every reader will be familiar with the conventions of chemical drawing, and some readers may choose to ignore the chemical structures presented in this and future chapters. In organic molecules such as those depicted in Figure 2.1, carbon atoms are assumed to be present at the intersections of lines, and enough hydrogen atoms necessary to give carbon four bonds are assumed to be present. For a more detailed understanding of molecular drawings, please see the Appendix.

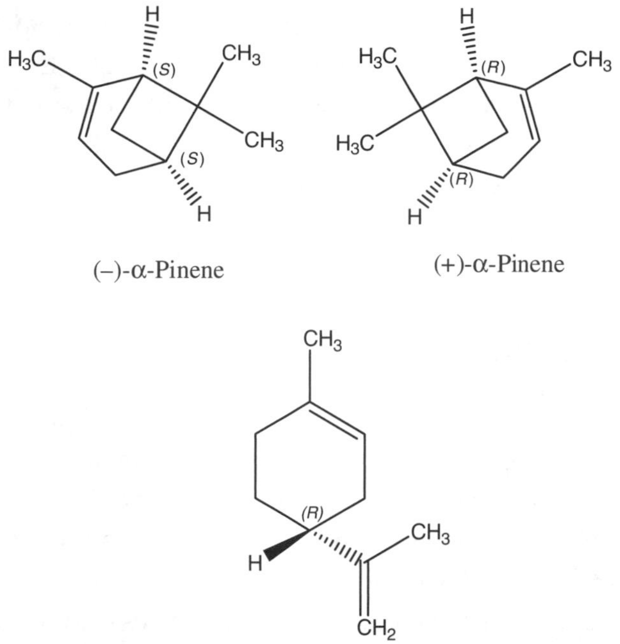

(−)-α-Pinene   (+)-α-Pinene

d-Limonene,
(+)-limonene

FIGURE 2.1. Molecular structures of α-pinene enantiomers and d-limonene.

Sidebar 2.A

Biot and Turpentine. The experimental measurement of the optical rotation of a flammable substance such as turpentine in the gas phase presented a formidable challenge for Biot and his coworkers. They set up the experiment in the cloisters of an ancient church, and constructed a tube 30 meters long in order to get enough molecules of turpentine in the gas phase to see a measurable effect. Of course, they also had to heat the turpentine with an open fire to produce turpentine vapor. Apparently, soon after they made the first key measurement, one of the wooden beams caught fire, and they had to call in some extra help to put out the flames. The experimental setup and the description of this accident are given in the seminal book on optical rotation by Lowry [1], originally published in 1933. As with other legends, this story has been embellished by many, and I have heard several lectures in which this story ends with the church burning down!

OPTICAL ROTATION

So why do chiral molecules rotate the plane of polarization? A very simple classical physics explanation is that the effect of an electric field is to accelerate charged particles in the direction of the field. A molecule can be considered as a collection of charged particles (electrons and nuclei). If a molecule has no plane of symmetry, then the charged particles on average must be accelerated by the field in a partially circular path. Circulation of charge leads to a magnetic force that interacts with the electromagnetic force of the applied field and results in a slight shift of the electric vector away from the initial plane of polarization. A very different but fundamentally equivalent explanation of this phenomenon was given by the brilliant French physicist Augustin Fresnel (Biographic Photo 2.2) in 1822, and his explanation is still used today when introducing this topic to beginning students in organic chemistry.

Fresnel realized that a linearly or plane-polarized light beam can be described by a sinusoidally varying electric field as plotted in Figure 2.2. The arrows (vectors) are used to indicate the direction and magnitude of the electric field ($\mathbf{E}$) of the light beam, which is assumed to be traveling in this figure from left to right (z direction). This light beam as drawn is x-polarized. An electric field is a *vector* quantity, which means that it has a direction in three-dimensional space (drawn as an arrow), and a magnitude that is depicted as the length of the arrow. Two

BIOGRAPHIC PHOTO 2.2. Augustin Fresnel.

vectors are added by positioning the tail of one vector on the head of another and drawing a new vector from the tail of the first to the head of the second as shown in Figure 2.3a. In this diagram we have illustrated the sum of vectors **A** and **B** to create the new vector **C**. An equivalent way of representing this vector addition is given in Figure 2.3b. Forces are vector quantities, for example, so application of pulling in directions **A** and **B** with

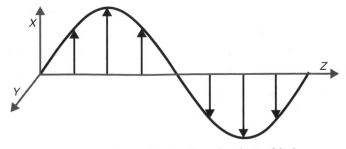

FIGURE 2.2. Linearly (or plane-) polarized light.

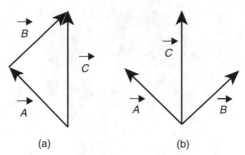

(a)   (b)

FIGURE 2.3. The addition of vectors.

equal magnitude is the same as pulling in direction **C** with the magnitude shown. Another way of stating this is that the vector quantity **C** may be *represented* by the two vectors **A** and **B**.

From the vector nature of light, it is not too difficult to see that the sinusoidally changing electric vector associated with a plane-polarized light beam can be represented as the vector sum of two circularly polarized light beams rotating in opposite directions as shown in Figure 2.4. In this figure the central arrow denotes the electric field of the plane- (or linearly) polarized light beam, and the left and right arrows within the circles represent the two opposite circularly polarized electric fields.

When the two circularly polarized beams travel through a medium composed of chiral molecules, they do so at different speeds. This is the key point in understanding this phenomenon. As the electric field rotates clockwise through a solution of chiral molecules, it meets a certain chiral distribution of charged particles (electrons and nuclei). It is the interaction with these charged particles that affects the speed through the medium. If the polarization is rotating in a counterclockwise direction through the same solution, the interactions will be different, and therefore the speed will be different. This is, perhaps, a bit easier to see if you think about a circularly polarized beam interacting with a right-handed or left-handed

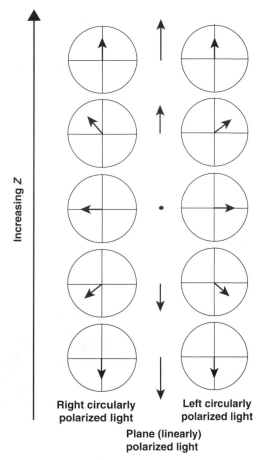

FIGURE 2.4. Linearly polarized light as a vector sum of left and right circularly polarized light.

helix, but the principle and effect are the same for all chiral molecules. So, if the speed of one of the circularly polarized components decreases relative to the speed of the other one, then, when these components exit the sample and "recombine" in a vector sense, the plane of linear polarization will be "rotated." This is illustrated in Figure 2.5.

Biot introduced the convention that a rotation of the plane of polarization in the clockwise direction to an observer viewing the beam as it is going away is said to be positive, and is

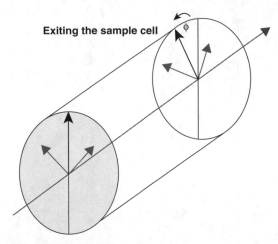

FIGURE 2.5. The rotation of plane-polarized light. The counterclockwise rotation shown in this figure is denoted as (−) and is called *levorotatory*. See color insert.

designated by (+) prefixed to the chemical name. A substance that induces a positive (clockwise) rotation is also said to be *dextrorotatory*. A substance inducing a negative (counterclockwise) rotation is said to be *levorotatory*, and is designated by the prefix (−). An old nomenclature for chiral molecules is to use the lowercase letters *d* and *l* as a prefix to indicate the direction of the rotation, for example, *d*-limonene is dextrorotatory limonene.

We should also note that the amount of rotation is a function of the wavelength of light used for the measurement. In the early experiments by Biot his source of light was simply the white light reflected off of bright clouds. In some later experiments he was able to study the effect of different wavelengths by using some available prisms. Biot noticed that in quartz crystals (which occur in chiral forms; see below) different colors (wavelengths) were rotated by different amounts. The empirical relationship that he determined is known as *Biot's law* and says that the amount of rotation α is inversely proportional to the square of the wavelength λ:

$$\alpha = \frac{k}{\lambda^2} \qquad \text{(Biot's law)} \qquad (2.1)$$

We now know that this is only an approximation, which is fairly accurate in the visible region of the spectrum. Optical rotation is widely used as a characteristic property of a chiral molecule, but in order to do so we need to specify a wavelength for the measurement. Historically, after the invention of the Bunsen burner in 1866, the orange light emitted by sodium chloride when placed in a flame was a good source of mono-chromatic light. Beginning in the early twentieth century, light from a sodium vapor lamp at 589.3 nm (the so-called sodium D line) has been used as a standard wavelength. However, if the molecule of interest absorbs this wavelength, a different color (wavelength) should be used. Synthetic chemists trying to produce a known chiral molecule will routinely determine the optical rotation of the substance as a measure of purity. For the species of interest one typically determines the *specific rotation*, which eliminates the dependence on cell length and concentration. The specific rotation is defined for a solution as follows

$$\alpha_{589.3}^{T} = \frac{\alpha}{l \cdot c} \qquad (2.2)$$

where α is the measured rotation in degrees, T is the temperature, l is the path length in decimeters, and c is the concentration in grams per milliliter. For a pure substance, c becomes the density. For the chiral compounds introduced above we can find the following data:

$\alpha_{589.3}^{20°C} = -51.28°$ for $(-)$-α-pinene (National Library of Medicine)

$\alpha_{589.3}^{20°C} = +115.5 \pm 0.1°$ for $(+)$-limonene in ethanol (Sigma Aldrich)

FROM BIOT TO PASTEUR

Some of the very first measurements of optical rotation were actually from quartz crystals, which were known by Biot and others to exist in mirror-image forms as shown in Figure 2.6. It was discovered that when the quartz was dissolved or melted, the optical rotation vanished. Louis Pasteur (Biographic Photo 2.3) compared this phenomenon to a set of spiral stairs that when dismantled no longer was chiral. He correctly attributed the asymmetry to the structure of the crystal, not the constituents. We now know that quartz is a crystalline form of SiO_2. Crystals are solids that have a regular repeating structure, and in the case of quartz, although the ratio of O atoms to Si atoms is 2 : 1, the basic repeating structure is tetrahedral Si atoms connected to four oxygen atoms. In quartz tetrahedral SiO_4 units can be arranged in different ways to give a repeating structure. The drawings in Figure 2.6 depict two mirror-image forms of what is called α-(low)-quartz. The macroscopic crystalline forms are, of course, connected to particular microscopic arrangements of the constituent atoms. In these structures the SiO_4 tetrahedra are situated in a helix along one direction in which each tetrahedron is turned by 120° relative to the previous

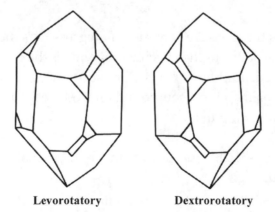

Levorotatory   **Dextrorotatory**

FIGURE 2.6. The two enantiomeric crystal forms of α-(low)-quartz.

BIOGRAPHIC PHOTO 2.3. Louis Pasteur. (Permission for use granted from the Edgar Fahs Smith Collection at the University of Pennsylvania Library.)

one, and the unit repeats with every fourth tetrahedron. In Figure 2.7 we show a segment of SiO_2 chains and on the left and right, the arrangement of the helical chains looking down the axis of the helix. We have tried to indicate depth by decreasing the font size of the atoms shown. In this figure, one has to imagine a very complex interconnected structure with every Si attached to four O atoms, and every O atom attached to two Si atoms. Clearly, Pasteur was correct in stating that the molecular building blocks of this structure are not chiral, but simply arrange themselves in a chiral (helical) form. This situation in which achiral units form chiral crystals is actually quite rare.

Biot also measured the optical rotation of various liquid substances and a number of naturally occurring substances dissolved in water. One of them was the chiral compound tartaric acid shown in Figure 2.8. This substance is obtained from *cream of tartar*, which is a residue that occurs in wine casks during the fermentation of grapes. Another substance

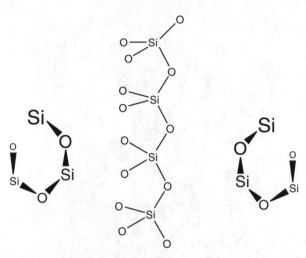

FIGURE 2.7. The arrangement of Si–O chains in α-(low)-quartz.

isolated from the tartar in some wine vats was called *racemic acid* from *racemes*, meaning bunch of grapes. This substance, when dissolved in solution, had the same physical properties found in tartaric acid but zero optical rotation. Pasteur's most important studies involving optical rotation were performed on tartrate salts. In particular, Pasteur grew crystals of the sodium (Na^+) ammonium (NH_4^+) salt of tartaric acid.

The salt can be formed by mixing the "free" tartaric acid with the bases sodium hydroxide (NaOH) and ammonium hydroxide ($NH_4OH \equiv NH_3 \cdot H_2O$). Note that the hydrogens (protons) are removed but the chirality of the tetrahedral carbons is unchanged. The simple chemical reaction describing

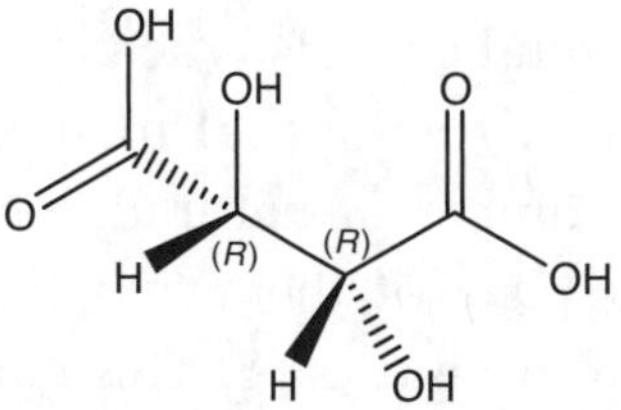

FIGURE 2.8. The structure and optical rotation of (+)-tartaric acid $\left(\alpha_{589.3}^{20^\circ C} = +13.5 \pm 0.5^\circ\right)$.

the formation of this tartrate salt is as follows:

$$NaOH \quad + \quad NH_4OH \quad + C_4H_6O_6 \quad \rightarrow$$
sodium hydroxide   ammonium hydroxide   tartaric acid

$$NaNH_4C_4H_4O_6 \quad + 2H_2O$$
sodium ammonium tartrate   water

Unlike the situation with the melting of quartz, when crystals of sodium ammonium tartrate were dissolved in water, they also rotated the plane of polarization. This led Pasteur to the correct conclusion that in this case, and many others, the phenomenon of optical rotation was a property of the molecules themselves, and not their arrangement in the crystal.

One of Pasteur's most important observations occurred while he was growing crystals of sodium ammonium salts of the "racemic acid" from an aqueous solution of this substance. Pasteur noticed that two different mirror-image crystals were formed. The ideal crystal shapes are shown in Figure 2.9. By carefully separating the two forms by hand and then redissolving them, he discovered that they rotated the plane of polarized light in equal and opposite directions. This led to the conclusion that racemic acid was a mixture of exactly equal parts of (+)-tartaric acid and (−)-tartaric acid. An exactly 50 : 50

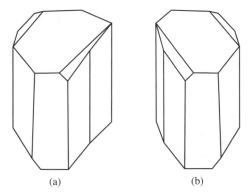

(a)   (b)

FIGURE 2.9. Crystal forms of (−) (a) and (+) (b) sodium ammonium tartrate.

mixture of two mirror-image forms of a molecular compound is now called a *racemic mixture* (see Sidebar 2.B).

Sidebar 2.B

Biot and Pasteur. Biot realized the importance of Pasteur's measurements. In fact, they were so important that he demanded that Pasteur repeat them in his (Biot's) laboratory using his equipment and his chemicals, and under his supervision. Biot insisted that the solution containing the two forms of tartaric acid be left under his observation while the crystals were formed. He let Pasteur separate the crystals, but he made up the two new solutions and made the optical rotation measurements himself. At this time (1848) Biot was 74 and Louis Pasteur was only 26. For Biot, tartaric acid rotated the plane of polarization to the right, but Pasteur demonstrated how he was able to have a form of tartaric acid rotate the plane of polarization to the left. When Biot saw the results, he is quoted as saying (in French, of course) "My dear child, I have loved science so much throughout my life that this makes my heart throb." He is said to have then given Pasteur his magnificent polarimeter, which is now located in the museum of the Pasteur Institute in Paris. Biot also became Pasteur's friend, advisor, and strong supporter. He helped him obtain various faculty positions throughout France, but unfortunately died 10 months before Pasteur's election into the Académie des Sciences of France in 1862. A very readable account of this experiment is given in an article by George B. Kaufmann and Robin D. Meyers [13]. The authors have included Pasteur's account in French, and their own very readable translation.

So, how many racemic mixtures crystallize in enantiomeric forms? Not many! Approximately 250 racemic substances are known to crystallize as a conglomerate of enantiomers [2]. An important and interesting fact is that Pasteur was very fortunate to be working in northern France and not in the warmer climate of southern France. We now know that if he had conducted his crystal growing experiments at or above 27°C (81°F), he would have seen quite different crystals being formed. In fact, at this elevated temperature, crystals containing equal amounts of the two mirror-image tartrate ions are formed but have very different crystal structures [3]. When these crystals are dissolved in a solution, there is no measurable optical rotation.

Although the original sample of racemic acid that Pasteur used for his experiments did come from wine vats, it was difficult to obtain more of this substance. The original source of racemic acid in Saxony was unable to reproduce the synthesis either because the particular source of the tartar was no longer available, or there was some accident in the purification that the chemists were unable to determine. At one point Pasteur traveled throughout central Europe sampling the tartars from various sources, but it appeared that new wine vats contained almost exclusively the (+)-tartaric acid. In 1857, the Société de Pharmacie de Paris offered a prize to anyone who could figure out a method to produce racemic acid. This prize was won by Pasteur when he showed that he could convert (+)-tartaric acid to the racemic form by mixing it with a substance named *cinchonidine* and heating it for more than 6 hours. The net result of this procedure was termed *racemization* of the tartaric acid. It became clear that the "natural" form of tartaric acid was (+)-tartaric acid.

You may be wondering why Pasteur was mixing his tartaric acid with cinchonidine. Cinchonidine is a chiral compound that is extracted from the bark of the red cinchona tree along with quinine, and was a well known substance in the midnineteenth century. In later years, Pasteur had great interest and success in using his chemistry knowledge to cure diseases, and no doubt substances like cinchonidine and quinine were of interest for their medicinal value. We now know that the structures of cinchonidine and quinine (Figure 2.10) are very closely related, and they both rotate the plane of polarization to the left. Pasteur actually had much success in separating racemic mixtures of tartrate salts by mixing them with chiral compounds such as cinchonidine and quinine and forming binary compounds know as *diastereomers* or *diastereoisomers* (see Sidebar 2.C). He first added acid (H^+) to form a positively charged ion from (−)-cinchonidine and then added the racemic acid. When he let the solution set and cool and then evaporate slowly, the first crystals that were formed were (−)-cinchonidine-(−)-tartrate. Eventually crystals of the diastereomeric complex (−)-cincho-nidine-(+)-tartrate were formed. The particular diastereomer-ic crystals could be redissolved and treated with base, and enantiomerically pure samples were obtained. Pasteur's inter-pretation of this separation procedure (which was clearly correct) was that one of the diastereomeric salts was more soluble than the other. Chemists refer to this process of sepa-rating racemic mixtures into individual enantiomers as *resolu-tion*. The method just described is still used today, and prior to very recent advances in chiral chromatography, which will

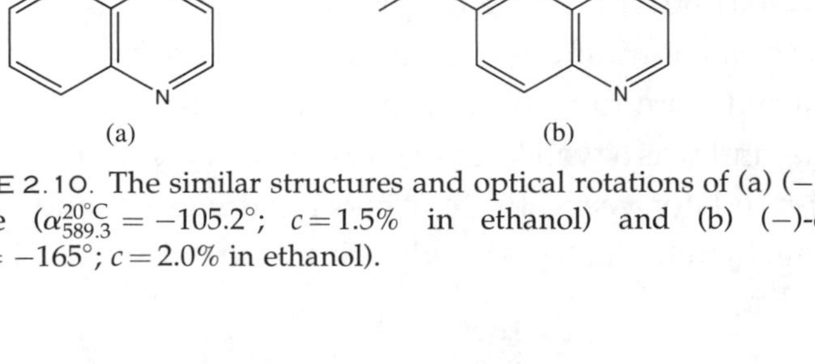

(a)   (b)

FIGURE 2.10. The similar structures and optical rotations of (a) (−)-cinch-onidine ($\alpha_{589.3}^{20°C} = -105.2°$; $c = 1.5\%$ in ethanol) and (b) (−)-quinine $\alpha_{589.3}^{20°C} = -165°$; $c = 2.0\%$ in ethanol).

be discussed later, chemists usually resolved enantiomeric mixtures in this way.

Sidebar 2.C

> **Human Diastereomers.** Binary compounds formed from individual enantiomers have very different properties, due to the differences in the interaction between, say, R–R and R–S. This is sometimes illustrated to students by having two of them shake hands (right–right) or hold hands (right–left). This demonstration always leads to differences in the "interaction" between the two students!

Pasteur's other very key experiment in the area of molecular chirality was to show that (−)-tartaric acid not only did not originate directly from the wine fermentation process but also did not participate in the biochemistry of fermentation. The simple experiments he performed involved adding some organic nutrients to an aqueous solution containing (+)-tartaric acid, letting it sit out on a warm day, and comparing what happened to a similar solution containing racemic tartaric acid (racemic acid). Both solutions were observed to ferment in response to the action of molds such as *Penicillium*, but the solution containing the "racemic acid" became increasingly left-rotating until the fermentation stopped and the rotation reached a maximum. What happens here is that all of the biologically "active" (+)-tartaric acid is used up in the fermentation process (it is food for the molds), and the solution becomes increasingly enriched in the biologically "inactive" (−)-tartaric acid. Clearly, the biochemistry of life as embodied in the fermentation process is one that has strict chiral selectivity! This property is often referred to as *homochirality*.

Perhaps we should finish the story about tartaric acid before going any further. Pasteur, of course, recognized the importance of molecular structure in three dimensions, and, in particular, the importance of molecular dissymmetry. Obviously, he didn't have the modern tools that we now have to determine the exact molecular formula and the exact three-dimensional structure. Perkin in 1867 was able to determine the basic chemical structure of tartaric acid as a four-carbon chain containing two hydroxyl (OH) groups and two carboxylic acid (CO_2H) groups [4]. Seven years later (1874) Le Bel published his finding on the connection between tetrahedral carbon and dissymmetry [5]. The structure of tartaric acid was finally confirmed by X-ray crystallography in 1923 by Astbury [6], although at this early stage of this technique it was not possible to determine the absolute configuration of the chiral centers. This was finally accomplished in 1951 by the Dutch crystallographers Bijvoet, Peerdeman, and von Bommel [7].

BIOCHEMICAL HOMOCHIRALITY

The realization that compounds such as tartaric acid had a dissymmetric structure led Emil Fischer (Biographic photo 2.4) in 1891 to propose the nomenclature system based on (+)-glyceraldehyde that was described in Chapter 1. In a truly amazing series of experiments, which eventually was recognized with a Nobel Prize in Chemistry, Fischer was able to determine the chemical formulas and relative structure of numerous sugars and amino acids. This was accomplished mostly by performing chemical transformations relating structures to each other and ultimately to (+)-glyceraldehyde. Of course, Fischer didn't know the structure of the enantiomer of glyceraldehyde that gave a positive rotation, so he just picked one of the two possibilities. We have these definitions drawn first as a two-dimensional "Fischer" projection and then with some indication of a three dimensional structure in Figure 2.11 As mentioned in Chapter 1, when the crystal structure of (+)-

BIOGRAPHIC PHOTO 2.4. Emil Fischer. (Permission for use granted from the Edgar Fahs Smith Collection at the University of Pennsylvania Library.)

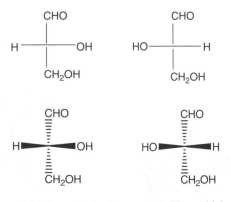

D-(+)-Glyceraldehyde   L-(–)-Glyceraldehyde

FIGURE 2.11. The molecular structures of the enantiomers of glyceraldehydes.

tartrate was determined more than 50 years later, one could use logic from the chemical transformations of Fischer to arrive at the conclusion that Fischer had guessed correctly.

Within months of the appearance of the article by the Dutch crystallographers Bijvoet, Peerdeman, and von Bommel on the absolute structure of (+)-tartrate, Linus Pauling (Biographic Photo 2.5), Robert Corey, and Herman Branson published a milestone article on the helical structure of proteins [8]. Fischer had actually shown 50 years earlier that proteins were a linear chain of amino acids, but the overall three-dimensional structure of proteins was unknown. We introduced L-amino acids in Chapter 1 in our discussion of nomenclature, and we draw a

BIOGRAPHIC PHOTO 2.5. Linus Pauling. (Public-domain photograph from the National Institutes of Health.)

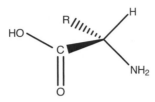

FIGURE 2.12. The structure of an L-amino acid.

three dimensional representation in Figure 2.12. The R in this structure denotes a generic chemical fragment or residue. We now know that almost all living systems on earth use only 20 L-amino acids (20 different R fragments) in the formation of the essential proteins of life.

There were a number of previous experimental measurements involving the diffraction of X rays from proteins that led to the conclusion that the structure of proteins was spiral (later called *helical*), and the major result of the Pauling work was to show how a stable helix could be formed from an amino acid chain. It is interesting to note that although some colleagues of Linus Pauling at the California Institute of Technology had heard about the X-ray crystallographic result on the structure of tartaric acid, and the implication about the absolute structure of L-amino acids, no one told him, or perhaps he didn't realize, the significance of this work [9]. Whatever the reason, when Pauling, Corey, and Branson published their paper, their diagram of the helix had the incorrect structure for the L-amino acids, and as a result Pauling's helix was left-handed. As we pointed out in Chapter 1, proteins form a right-handed helical structure. Fischer guessed correctly, but Pauling guessed wrong.

In 1952, M. L. Huggins [10] calculated the interatomic distances for helical polypeptide chains and showed that one couldn't make a left-handed helix out of L-amino acids that fit the experimental X-ray diffraction data because a carbon from one turn came too close to an oxygen on the next turn. However, a right-handed helix made from L-amino acids (or a left-handed helix made from nonnaturally occurring D-amino acids) leads to an extra stable structure as a result of attractive interactions (so-called hydrogen bonding) between the turns.

So we now know that proteins, which constitute about 15% of the mass of the human body, are composed of only L-amino acids. L-Amino acids represent about 40% of our "dry mass" (our body is approximately 65% water), and the consequences of having L-amino acids are that the alpha helices are right-handed. Other larger scale three-dimensional structures are seen for proteins, but they are all the result of life on this planet using only one of the two possible mirror-image forms of amino acids.

Other major constituents of life are sugars. A simple sugar (or monosaccharide) is composed of five to eight carbon atoms with the type of structure given in Figure 2.13. This is actually the Fischer projection of D-glucose; the most common sugar that Fischer knew as "grape sugar" is now often referred to as "blood sugar." Remember that in a "Fischer projection" the horizontal bonds are assumed to be above the plane of the paper, and the vertical lines are intended to represent bonds that point from the central carbon to the back of the paper. The actual structure of sugars in solution is very complex and variable. D-Glucose exists in aqueous solution as an interconverting mixture of cyclic forms in which the two ends of the molecule are connected together. An attempt at a three-dimensional structure representing one of these cyclic forms is given in Figure 2.14. It might take some mental gymnastics to convince yourself that Figures 2.13 and 2.14 depict the same molecule!

This situation is quite different from what we saw with L-amino acids because in that system we had only one chiral carbon center. For glucose we have four chiral centers, and since each can be either R or S, we have $2 \times 2 \times 2 \times 2 = 16$ possible

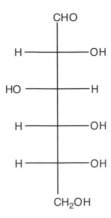

FIGURE 2.13. Fischer projection of D-glucose.

chiral structures. Only one of these structures is the naturally occurring glucose as drawn. Emil Fischer chose the bottom chiral carbon shown in Figure 2.13 for his nomenclature, so comparison of the structures in Figure 2.11 with that in Figure 2.14 shows that this molecule is, indeed, D-glucose. In L-glucose all of the chiral carbons are inverted, and the other possible structures with different combinations of R and S chiral centers all have different root names with prefixes D or L. For example, the structure depicted in Figure 2.15 is the sugar D-mannose, which is found in some fruits, especially cranberries. It turns out that almost all naturally occurring sugars are

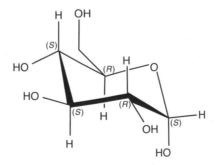

FIGURE 2.14. Molecular structure of the α-pyranose form of D-glucose.

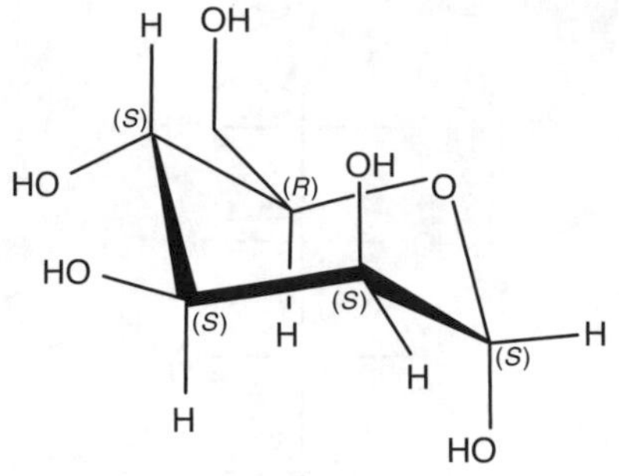

FIGURE 2.15. Molecular structure of the α-pyranose form of L-mannose.

D-sugars. This therefore is another example in which living substances display chiral selectivity.

The exact "absolute" structure of many sugars was also determined after the structure of (+)-tartaric acid was determined by Bijvoet, Peerdeman, and van Bommel. Two other very important naturally occurring sugars containing five carbon atoms (pentoses) and three chiral carbon centers are D-ribose and 2-deoxy-D-ribose. These are drawn in Figure 2.16. The difference in these two sugars is the removal of an oxygen ("deoxy") from the carbon in D-ribose, which chemists designate by convention as carbon 2. When these two sugars exist as cyclic species in a long polymer chain, connected by phosphate (PO_4^{3-}) linkages, and combined with certain so-called purine and pyrimidine bases, they exist in the nucleus of every living cell and are collectively termed *ribonucleic acid* (RNA) or deoxyribonucleic acid (DNA).

The three-dimensional structure of DNA was reported in a landmark paper by James Watson and Francis Crick in 1953 [11]. As in the determination of protein structure, certain experimental X-ray diffraction data indicated that the overall structure was helical. Earlier that same year, Linus Pauling and Robert Corey published a proposed structure for DNA that was composed of three intertwined chains [12]. Unlike the Pauling–Corey model, the Watson–Crick intertwined "double

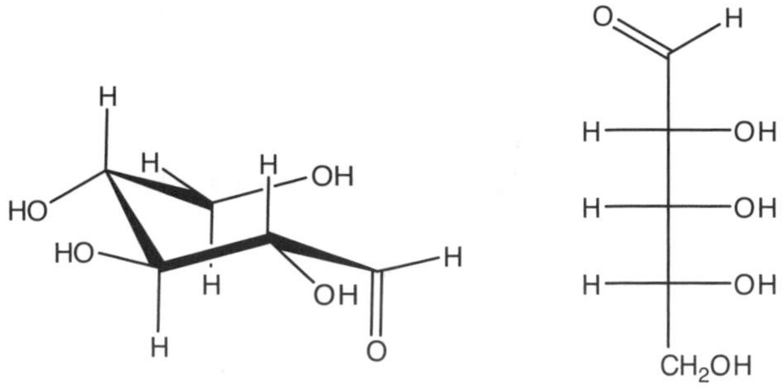

D-Ribose

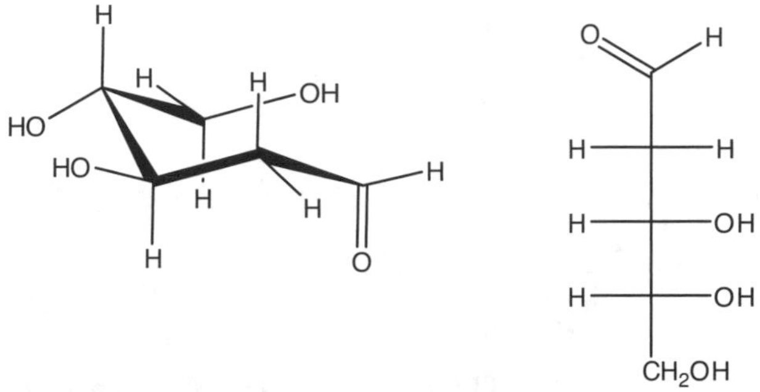

2-Deoxy-D-ribose

FIGURE 2.16. Molecular structures and Fischer projections of D-ribose and 2-deoxy-D-ribose.

helix" was able to explain all the existing data, and also made perfect sense chemically in terms of interatomic distances and attractive hydrogen-bonding interactions. This discovery was made 2 years after it was determined that Fischer got it right with the structure of (+)-glyceraldehyde, so Watson and Crick knew the absolute structure of D-ribose and 2-deoxy-D-ribose. When they built their models from these chiral building blocks, they also got a right-handed helix (actually a double right-handed helix) as shown in their original diagrams and physical models. A segment of double-helix DNA is shown in Figure 2.17 In this diagram the helical chains, which are formed from alternating deoxyribose and phosphate (PO_4^{3-}) ions, are indicated by colored "ribbons" in order to see the right-handed

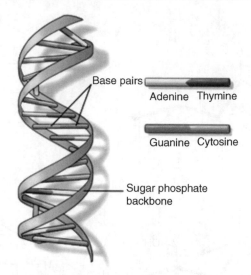

Base pairs

Adenine   Thymine

Guanine   Cytosine

Sugar phosphate
backbone

FIGURE 2.17. A three-dimensional model of a section of DNA. (Public-domain diagram from the US National Library of Medicine in the National Institutes of Health.) See color insert.

double-helix structure. The two separate helices are connected by specific achiral base pairs as shown. RNA exists as a single or double right-handed helix, and the reader is referred to the references at the end of this chapter for an introduction to the biological functions of these very important molecules.

DETERMINATION OF ABSOLUTE STRUCTURES OF CHIRAL MOLECULES

Although the measurement of optical rotation is an extremely important tool in helping chemists determine whether a chiral substance is pure, the technique has not lent itself to reliable theoretical modeling. In other words, even with the very sophisticated understanding we have today of the electronic structure of molecules, we are not yet able to reliably predict an optical rotation from a given absolute structure. This is true not only for the magnitude of the rotation but also for the sign of the rotation. So, even if we know that a compound is pure, and have a good value for the optical rotation from a careful experimental measurement and knowledge of the basic chemical structure (except for the identity of the enantiomer), we cannot yet determine with certainty which enantiomer we have without some additional structural information. This additional information may be, for example, results from a compound with a very similar structure for which we know the rotation and absolute structure, but even these cases may lead to errors in identification of chirality.

There are other spectroscopic methods for probing the structure of chiral molecules. All chiral molecules will display differences in the scattering, absorption, or emission of circularly polarized light, but unfortunately the differences tend to be quite small. Whereas optical rotation involves the scattering of circularly polarized light, *circular dichroism* (CD) is the difference in absorption of left circularly polarized light versus right circularly polarized light. In CD one scans the wavelength region corresponding to a particular absorption for the chiral molecule in question. Instrumentation for this measurement has been available since the 1960s, and although extremely important structural information has been obtained using CD in the ultraviolet and visible spectral regions since the 1960s, there has been only limited success in predicting the absolute identity of a chiral molecule without additional knowledge. All chiral molecules that emit light (fluorescence or phosphorescence) also will show a preference for right or left circular polarization. This technique, referred to as *circularly polarized luminescence* (CPL), has shown wide applications as a probe of the structure and dynamics of luminescent chiral molecules, but also has not yielded very much information on absolute structure.

There have been some very promising results in the use of CD in the infrared region. Infrared radiation excites molecular vibrations, and this technique is called *vibrational circular dichroism* (VCD). Comparing VCD measurements of certain classes of molecules with the predictions of very modern calculations has shown this technique to be quite helpful in determining absolute structure.

The most reliable and applicable technique remains single-crystal X-ray crystallography. Not every molecule is available in sufficient quantity and purity to grow suitable crystals, and unfortunately not every chiral molecule crystallizes. The use of X-ray diffraction developed by Bijvoet relies on a phenomenon called the *anomalous dispersion of X rays*. When the wavelength of the X ray is close to an absorption band of an electron on an atom in the crystal, an additional phase shift is imposed on the scattered X-ray beam. This phase shift has on effect on the overall interference pattern, allowing the X-ray crystallographer to determine not only the distance between a pair of atoms in the crystal but also whether a particular atom is in front of or behind the scattering center. Improvements in instrumentation and theoretical analysis have led to an expansion in the number of X-ray determinations of absolute structures, but the technique is difficult to apply to molecules containing only light atoms because the electronic structure of the heavier atoms make them more suited to the X-ray absorption required [2].

SUMMARY

In this chapter we have described the chronology of discoveries that has led to the conclusion that life is chiral. The facts that amino acids, which constitute almost all of living material, are all L, and that the sugars that form the helical backbone of all living DNA and RNA are D, were well known by the midtwentieth century. An obvious question or two arise. Must the building blocks of life be chiral for life to exist? Could there be life based on racemic building blocks, or perhaps achiral molecules? These questions cannot be answered with absolute certainty, but it is possible to make some general conclusions about this aspect of life. These are the topics of the next chapter.

SUGGESTIONS FOR FURTHER READING

Watson, J. D., *The Double Helix: A Personal Account of the Discovery of the Structure of DNA*, Touchstone, New York, 1996.

Serafini, A., *Linus Pauling: A Man and his Science*, Paragon House, New York, 1991.

Debré, P. and E. Forster, *Louis Pasteur*, Engl. transl., Johns Hopkins Univ. Press, Baltimore, 1998.

Mason, S. F., *Chemical Evolution: Origin of the Elements, Molecules and Living Systems*, Clarendon Press, Oxford, 1991.

Charney, E., *The Molecular Basis of Optical Activity: Optical Rotatory Dispersion and Circular Dichroism*, Wiley, New York, 1985.

REFERENCES

1. Lowry, T. M., *Optical Rotary Power*, New York, Dover, Mineola, NY, 1964.

2. Mason, S. F., *Molecular Optical Activity and the Chiral Discriminations*, Cambridge Univ. Press, Cambridge, UK, 1982.

3. Scacchi, A., *Atti Accad. Sci. Napoli* **4**:250 (1865).

4. Perkin, W. H., On the basicity of tartaric acid, *J. Chem. Soc.* **20**:138–176 (1867).

5. La Bel, J. A., On the relations which exist between the atomic formulas of organic compounds and the rotatory power of their solutions, *Bull. Soc. Chim.* **22**:337–347 (1874).

6. Astbury, W. T., The crystalline structure and properties of tartaric acid, *Proc. Roy. Soc. A* **102**:506 (1923).

7. Bijvoet, J. M., A. F. Peerdeman, and A. K. von Bommel, Determination of the absolute configuration of optically active compounds by means of X-rays, *Nature* **168**:271 (1951).

8. Pauling, L., R. B. Corey, and H. R. Branson, The structure of proteins: Two hydrogen-bonded helical configurations of the polypeptide chain, *Proc. Natl. Acad. Sci. USA* **37**:205–211 (1951).

9. Dunitz, J. D., Pauling's left-handed α-helix, *Angew. Chem. Int. Ed.* **40**:4167–4173 (2001).

10. Huggins, M. L., Polypeptide helixes in proteins, *J. Am. Chem. Soc.* **74**:3963–3964 (1952).

11. Watson, J. D. and F. H. C. Crick, A structure for deoxyribode nucleic acid, *Nature* **171** (3):737–738 (1953).

12. Pauling, L. and R. B. Corey, Structure of the nucleic acids, *Nature* **171**:346 (1953).

13. Kaufmann, G. B. and R. D. Meyers, Pasteur's resolution of racemic acid: A sesquicentennial retrospect and a new translation, *Chem. Educator* **8** (3):1–18 (1998).

LIST OF BIOGRAPHIC PHOTOGRAPHS, SIDEBARS, AND FIGURES

THE ORIGIN OF CHIRALITY IN LIVING SYSTEMS

It should now be evident that the most important molecules that exist in the cells of all living things on earth are almost all chiral. For more than 100 years scientists have known that the building blocks of proteins and DNA are composed of only one of the two mirror-image forms of amino acids and the special group of sugars, and a little more than 50 years ago precisely which mirror-image form was determined. We have also seen homochirality when these chiral building blocks are used to form the much larger helical structures present in proteins and DNA. In the forthcoming chapters we will discuss the consequences of these chiral molecular structures on our macroscopic world, but in this chapter we focus on what is commonly referred to as the "origin" of chirality in our living world at the microscopic molecular level.

Three obvious questions follow from the observations presented in Chapter 2:

1. Does life have to be chiral, or could life develop that uses, for example, a racemic mixture of amino acids?

Mirror-Image Asymmetry: An Introduction to the Origin and Consequences of Chirality by James P. Riehl
Copyright © 2010 John Wiley & Sons, Inc.

2. Could life develop using D-amino acids and L-sugars?

3. If the answer to question 2 is "yes," then why, indeed, does life on earth use L-amino acids and D-sugars, and not D-amino acids and L-sugars?

These are the three principal topics of this chapter.

MUST LIFE BE CHIRAL?

It is, perhaps, useful to think about the operation and construction of a complicated mechanical device such as an automobile as we ponder the even more complicated biomechanical living system. Looking at a car moving along a road from a distant overlook, we would see something that appears to have bilateral symmetry. The car has a front and a back that are different, and a top and bottom that are different, but at this distance the one side resembles the mirror image of the other side. From this vantage point the car does not appear to be chiral. If we get a closer look at the car, however, we see that the driver is holding a steering wheel on one side; there are no planes of symmetry, so the car is in fact chiral. In fact, every car we see has the steering wheel on the same side, so we come to the conclusion that automobiles all, in fact, have identical chirality.

A little traveling to the U.K, Japan, Australia, and a number of other countries, however, might be disconcerting because the cars look the same at a distance, but a closer look shows the steering wheel to be on the opposite side of than seen in the United States or continental Europe. (see also Chapter 8). So, what can we speculate about the construction of these automobiles? What are they made of? Of course, at the factory we see that automobiles do not have bilateral symmetry. This is easily seen after opening the hood, but it is true inside the car as well because of the steering wheel, glove compartment, and other components. So our next stop is the parts department, where we will surely see shelf after shelf of chiral automobile parts and, of course, that the smallest parts such as screws are chiral, and, in fact almost are right-hand-threaded. Suppose that the parts department had an equal mixture of all right-handed parts such as screws and an equal number of left-handed parts and screws. Would this make the job of the auto mechanic or auto assembler any easier? Well, of course, not! For many reasons it just makes sense for all the screws to be designed for either right- or left-hand use. Very rarely, situations arise calling for screws designed for use of the opposite hand, but these special cases need to be pointed out to the mechanic, who would otherwise be struggling and struggling to unscrew a part only to be in fact making it tighter and tighter. Perhaps it is time to leave the automobile analogy, but hopefully you get the point that for things as complex as an automobile or a living cell, it just makes sense for all the building blocks to have the same chiral identity.

The chemistry of living things (biochemistry) is extraordinarily complicated, energy-efficient, and selective. Of course, life has had billions of years to develop these chemical processes, and it takes incredible efforts by scientists to unravel and understand even the simplest biochemical reaction. Even though chemists have made amazing advances in the last 100 years in the synthesis of very complicated molecules, they have come nowhere close to the selectivity and complexity that is inherent in the chemistry of living systems. Even some of the simplest chiral molecules, such as the L-amino acids, are made in pure form as needed by the living cell, but if we need to make them in the laboratory it is a daunting challenge, indeed. The traditional synthetic chemist is actually faced with two choices, starting the synthesis with some chiral material that most likely has a living origin (so-called natural product), or to produce the racemic mixture from an achiral starting material, and find an efficient way to separate the enantiomers. More recently, there have been remarkable advances in harnessing living microorganisms to do these syntheses for us.

However, we haven't yet answered the question about whether chiral exclusivity is necessary for life. Even the simplest understanding of a biochemical process leads one to the general conclusion that biochemistry must be three-dimensional. It is inconceivable that the necessary biochemical processes could employ flat molecules or other symmetric species. One of the underlying principles of life is recognition and response. Recognition at the level of sophistication required for life must be asymmetric. Furthermore, life must be energy-efficient. It makes no sense for a living system to expend the energy to produce racemic mixtures where there is use for only one enantiomer. Just as inefficient would be a living system that was functioning with two competing pathways. Should the screw manufacturer supplying our automobile factory produce both mirror-image screws, knowing that only one of them will be used? We know the answer, of course, for life in our world, and it just makes sense that all life in all worlds should display chiral exclusivity.

MIRROR-IMAGE LIFE?

Perhaps you are now convinced that for the variety of reasons given in the previous section that life anywhere must certainly be chiral, and, therefore, composed of chiral building blocks. Soon after the determination of the exact absolute structures of amino acids, the question that arose in the minds of many scientists is why there are L-amino acids and D-sugars, and whether life could be possible in a completely mirror-image world. (see Sidebar 3.A). At the molecular to macroscopic level, there doesn't appear to be any reason why mirror-image life couldn't exist. At this level there is no energy difference between mirror-image systems, although, as we will soon see, at the subatomic level there is an inherent chirality that we need to consider. When that first alien space ship arrives on our planet, there will surely be some scientists out there collecting samples to see if the amino acids (if this creature even uses amino acids!) are L or D.

Sidebar 3.A

Alice's Mirror-Image Milk. In Lewis Carroll's *Through the Looking Glass*, Alice tells her cat, Kitty, that the house she sees in the looking glass looks like the house they live in, but does worry that "Perhaps looking-glass milk isn't good to drink..." Perhaps, Alice was aware that milk does contain some D-amino acids. D-Amino acids are a component of bacteria cell walls, and must help protect the bacteria from enzymes that the L-amino acid–based host organism uses to protect itself from bacterial infection. Most of the D-amino acids in milk are thought to come from microorganisms that cows eat as part of their diet, not from bacteria present in fresh milk [16]. Aged cheese such as Gouda, Appenzeller, Parmesan, and Emmental contain much more D-amino acids than are present in milk. In this case, the D-amino acids come from the bacteria cultures added to make the cheese. In fact, the amount of D-amino acid can be used to monitor the aging of certain cheeses. In a nice full meal of raclette or fondue, a person might consume as much as 300 mg of D-amino acids. Alice need not worry, because it appears that consumption of D-amino acids at this level is not harmful.

THE ORIGIN OF HOMOCHIRALITY IN LIVING SYSTEMS

Why are all proteins based on L-amino acids and not D, and why do D-sugars predominate over L-sugars in living systems on earth? Before we try to answer this question, we need to know something about the beginnings of life on earth. Many experiments have been performed in which various guesses at the constituents of the primordial earth atmosphere were subjected to electric discharges and/or UV light to see what organic molecules might be formed. Stanley Miller, while working in the laboratory of Professor Harold Urey at the University of Chicago in the early 1950s, was the first to show that amino acids could be formed when mixtures of gases such as methane (CH_4), ammonia (NH_3), water (H_2O), and hydrogen (H_2) were subject to electric discharge [1,2]. Since the beginning of life on earth is presumed to have formed in the primordial oceans or pools, these gases were added to a system that also contained liquid water. In these very first experiments several chiral amino acids were formed in enough quantity to be identified, and many more were apparently formed in smaller quantities. It is important to note that these compounds were all formed as racemic mixtures. We will discuss this soon, but it should be evident that without the influence of an external "chiral" force, or without the influence of living chiral systems, the formation of these building blocks of life will lead to racemic mixtures. There has now been more than 50 years of research in this area, and it is clear that all of the important building blocks of life in their racemic form could be synthesized from "earth, air, fire, and water." It might be, of course, that the very beginnings of life did not rely on the precise compounds that we know about today, but it is reasonable to conclude that the actual sugars, amino acids, or their precursors that formed the basis of the very first self-replicating system could have formed in the same way.

So let's think about this pool of water containing this complex mixture of organic molecules. If we quite reasonably assume that water was in its liquid state, we have quite a narrow temperature range to consider, namely, 0–100°C. The primordial earth had billions and billions of pools and perhaps a billion years to find just the right conditions to get life started. One can speculate that all life on earth as we now know it sprang from the very rare chemical event when some molecular species replicated itself, and found a way to use some energy source to continue the process. This process as described is strictly terrestrial; that is, no extraterrestrial source of starting material was necessary. We now know, of course, that some very complex organic molecules are delivered to the earth on meteorites, and perhaps in the beginning of planet Earth asteroids or comets were the source of the building blocks of life. One recent estimate is that as much as 50% of the water on earth is the result of comet impacts [3].

EXTRATERRESTRIAL CHIRALITY

In order to learn about what compounds might be delivered to the earth it is useful to examine meteorites that have survived the fiery trajectory through the atmosphere. One of the most widely studied meteorites is the Murchison meteorite, which fell about 60 miles north of Adelaide, Australia in 1969. One of the most interesting and confusing studies on this meteorite concerns the amino acids that were found when the composition of the meteorite was analyzed. One of the first studies reported that a number of amino acids were detected, some among the 20 necessary for life on earth, and some that were "nonnatural" [4]. This first report also stated that the amino acids that were found were racemic. A subsequent study reported an excess of the L-amino acids, which would, of course be consistent with a conclusion that the source of our chirality might be extraterrestrial [5]. Another study appeared claiming that the second study was in error, and further papers followed dealing with the measurements and interpretation of the amino acid content of this meteorite [6]. One issue has been whether the meteorite has been contaminated through human handling or by the measuring devices. Concentration of the chiral organic molecules is very low in these extraterrestrial samples, so it wouldn't take much impurity to cause a significant contamination. What is needed, of course, is a way to collect a sample from an object from space that has not been contaminated by passing through the chiral earth's atmosphere and sitting on the chiral earth's surface.

From the most recent measurements that have been published on the Murchison meteorite, it does appear that certain amino acids are present in nonracemic enriched form. In fact, the results show that the form in excess is the L-isomer! Most of the enantiomeric excesses that have been reported in the literature are quite small; however, one new result shows excesses of 15% for one particular amino acid [6]. If we found only one enantiomer in an extraterrestrial sample that we were sure did not originate from earth, we would be almost forced to conclude that life is present somewhere in our universe. Although there are physical processes that could generate the destruction or formation of a specific enantiomer from a racemic mixture, exclusive homochirality would seem to be a unique aspect of life (at least life as we know it!). For example, the absorption of high-energy circularly polarized light by a molecule might lead to bond breaking and destruction of the molecule, and this would lead to an excess of one enantiomer over the other if the probability of absorption differs between the two enantiomers. This is rigorously true if the light is circularly polarized, and it is straightforward with modern instrumentation to measure the difference in the absorption probability for left and right circularly polarized UV light by amino acids or other chiral molecules. So exposure of a racemic organic mixture in outer space to high-energy circularly polarized radiation will yield an enantiomeric excess if one of the two enantiomers preferentially absorbs the radiation and is therefore preferentially destroyed, yielding an excess of the weaker absorbing species.

It is important to remember that both enantiomers are absorbing the radiation and being destroyed, one is just being photochemically destroyed more rapidly than the other. The longer the exposure to circularly polarized light, the more chiral the sample becomes, but, of course, too long an exposure will destroy all of the molecules. A rough estimate is that a chiral excess of 10% is probably the maximum one could expect from this process.

Astronomers believe that there are a number of sources of circularly polarized light in outer space [7]. Sunlight striking the earth, for example, is partially circularly polarized, due to the earth's magnetic field and reflections off clouds and other surfaces. In order to accomplish the level of chiral-selective photodestruction required to generate chiral molecules, however, one needs circularly polarized ultraviolet light, since this corresponds to the amount of energy usually required to break molecular bonds. Earth-bound instruments are unfortunately unable to detect circular polarization in the UV radiation that strikes the earth because of scattering in the atmosphere. There was a device on the Hubble telescope capable of measuring polarization (i.e., the "faint-object camera"). This particular instrument was seldom used, and eventually removed from the telescope in one of the space shuttle service missions. Astronomers do see circularly polarized light at longer wavelengths and have extrapolated these observations back to shorter UV regions. It is speculated that UV circular polarization is very likely generated by neutron stars, magnetic white dwarfs, reflection off nebulae, and other astronomical sources.

Are we to assume, then, that the universe is providing a chiral environment to molecules present in outer space, generating the excess enantiomers that we need to impinge on earth and start the self-replicating chemistry leading to life? From what we know of the universe and the physics of how circularly polarized light would be generated, one can envision regions of space in which one circularly polarized component would dominate and other regions in which the other would be in excess. It is certainly possible that our region of interstellar space contains the circular polarization necessary to generate the excess amino acids or other chiral molecules to get chiral life started, but as we have seen, there are so many unknowns that it is impossible to arrive at a conclusion to which all scientists would agree.

Although the most recent and presumably best analytical results from meteorites do indicate the presence of an excess of L-amino acids, the results obtained thus far are not conclusive for many scientists. Once a meteorite or dust particle enters our environment, it is exposed to a system in which L-amino acids dominate. The amount of amino acids found on meteorites is very small, and how can we be sure that that the system has not been contaminated? Perhaps the answer to this question will have to wait until the year 2014. This is when the robotic lander Philae will detach itself from the orbiter of the Rosetta space-craft and land on the comet 67P/Churyumov-Gerasimenko [8]. Some fairly sophisticated chemistry will be performed in order to separate and analyze the organic compounds that are dis-covered. One of these experiments will be to determine the chirality of the compounds that are detected. Maybe then we will know whether our region of space is producing an excess of L-amino acids. (see Sidebar 3.B).

Sidebar 3.B

> **The *Rosetta* Mission.** The Rosetta spacecraft with the lander Philae is the first spacecraft to include an experiment to detect chiral molecules outside our planet. The spacecraft was named in honor of the Rosetta stone, which allowed Egyptian hieroglyphics to be translated, and resulted in

learning about civilization 2300 years ago. The lander Philae was named for an island in the Nile River on which an obelisk was found that helped in translating the Rosetta stone. Hopefully Rosetta and Philae will give us a view of the universe 4.6 billion years ago when comets and earth were formed. Rather than measure optical rotation, which is too difficult to accomplish in the difficult environment of a comet surface, Philae will attempt to separate the enantiomers of any chiral molecules found by a technique known as *chiral chromatography* [17]. In this technique a vapor phase containing different enantiomers will be separated by passing through a tube containing a chiral solid material at different speeds. The compounds will be analyzed by determining their mass in a mass spectrometer. As of December 2007, Rosetta is on schedule to reach the comet in March 2014, and Philae is expected to land in November 2014.

(Copyright 2007, European Space Agency)

NUCLEAR CHIRALITY

From all we have discussed so far and from what we all experience in our part of the universe, it would be quite natural to conclude that in a universe that is a mirror image of the one we live in, that the laws of physics and chemistry would be unchanged. In the simple physics we see in our everyday lives of machines, objects falling, or thunderstorms forming, or in the larger world of planetary physics there is no observable difference between what we see and what would occur in a mirror-image version. The nuclear physicists among the readers of this book know, however, that this conclusion is incorrect for the physics of elementary particles. The property that is used to describe the effect of reflection in this case is known as *parity*. If everything remains the same on reflection, we would say that "parity is conserved"; if not, then parity is not conserved.

Actually, before 1956 physicists believed that parity was conserved in all systems, but in a remarkable story of theoretical and experimental breakthroughs described wonderfully by Martin Gardner in his book *The New Ambidextrous Universe* [9], it was speculated and then verified that parity was not conserved in the so-called weak nuclear interaction. The experimental evidence for this conclusion was provided by Chien-Shiung Wu from Columbia University and her coworkers [10], based on theoretical predictions of Yang and Lee [11]. This first experiment involved cooling radioactive cobalt-60 nuclei down to almost absolute zero, and applying a magnetic field so that most of these magnetic nuclei were aligned with the magnetic field. Most physicists expected at that time that when these radioactive nuclei decayed by ejecting an electron, that they would be ejected with equal probability along both magnetic axes. However, the result was that the electrons were more likely ejected out of the south end of a cobalt-60 nucleus than the north end. It is far beyond the scope of this book to explain these experiments or their meaning in the world of elementary particle physics in further detail, but it is clear that there is a difference between the north end and south end of a cobalt-60 nucleus. As Martin Gardner points out, the picture of a spinning sphere for this nucleus had be replaced by a spinning cone, although these are just ways to visualize the physics and are not meant to assign shape or motion to the nucleus. For our discussion we note that a spinning cone is chiral and a spinning sphere is not! (See Figure 3.1.)

   If atoms are now described as "chiral," then we have to reexamine what we mean by the term *molecular enantiomers*. The pictures given in Chapter 1 of molecules and their mirror images changed the spatial arrangement of the atoms, but left the nuclei unchanged. In our new picture we need to change particles into antiparticles as well as reflect the spatial arrangement of atoms. If we do only the spatial part, then the two sides of the mirror are not true enantiomers, and will not have identical energies. Calculation of the energy difference between

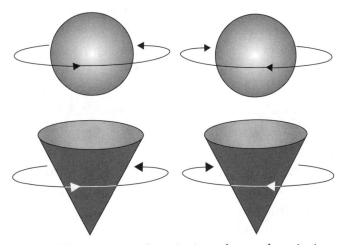

FIGURE 3.1. Mirror images of a spinning sphere and a spinning cone.

spatial enantiomers, due to the fact that the nuclei are "chiral," has been the target of numerous investigations.

One such calculation of the energy difference between spatial enantiomers of D- and L-amino acids has been reported by Tranter [12], who calculated the energy difference between D- and L-amino acids is to be 10^{-38} J, with a preference for the L-amino acid enantiomers. This is a very small energy difference, and if we use normal statistics to calculate how many D and L molecules would exist at room temperature simply on the basis of this energy difference, we calculate that in 10^{18} molecules (a million trillion molecules) there would be one more L-amino acid than D-amino acid due to this nuclear non-parity-conserving weak interaction. This is not very many! Although this difference is calculated to be very small, there have been numerous attempts to design and perform experiments to see if this effect could be measured. Unfortunately all these experiments have failed to produce results that illustrate the effect of this type of chiral energy discrimination [13].

OTHER EXTRATERRESTRIAL SOURCES

There are some other far-fetched and perhaps not so far-fetched ideas about the source of earth chirality. In recent years, there have been numerous reports about meteorites found in the arctic region that have been attributed to coming from Mars, and some of these reports presented evidence that was interpreted to indicate the existence of remnants of living microbes. It is not too difficult to imagine that if there were events on Mars that ejected Martian stuff out into space that found its way to planet Earth, perhaps some chiral molecules were deposited on our planet. These molecules could then have started the process of life on earth.

More to the liking of movie studios, perhaps, is the idea that we were visited in the distant past by extraterrestrial beings and that these beings left some "remains" on earth that evolved into us. Or maybe, as our astronauts do, while orbiting and surveying this unknown and at that time uninhabited planet, they just disposed of some "waste" that found its way down to the surface without being completely destroyed? There is certainly evidence for this in *Star Trek* and other science fiction movies where beings on other planets tend to resemble humans.

THE BEGINNING OF LIFE ON EARTH: BIOCHEMISTRY

Let us ignore for the remainder of this book the idea that we were visited by aliens, and return to a more scientific discussion of how chiral life began. We have seen in this chapter various theories as to how the primordial ponds could have been seeded with L-amino acids or their precursors. The two main ideas discussed so far are that chiral organic molecules rained down on the earth billions of years ago, or alternatively there is a natural energy difference between spatial enantiomers, resulting in a "natural" difference in the concentration of the different enantiomers that may have been present. In each of these cases, one predicts a mixture that is not racemic, but provides the chemical events to the beginning of life with an "enriched" chiral pool of starting material.

A natural question is how much of an excess do we need in the chiral pool. Do we need the 10% or so that the optimistic predictions of chiral photodestruction in space might give us, or will the one molecule in 10^{18} predicted by the parity-violating weak interaction be sufficient? In thinking about this question we need to keep in mind again that we have billions of pools over billions of years, so that even if the chemical reaction that we need to get life started is very rare, we have lots of time and many pots to work with. In most of the attempts to answer this question, scientists have assumed that the series of reactions that need to occur are *autocatalytic*, meaning that the product of the reaction influences the speed of the reaction that produced the product. So if A reacts with B to produce C at a certain rate, then the presence of C increases this rate. Perhaps you can see qualitatively how a reaction could "take off" since the more you produce, the faster the reaction goes. In a competition, albeit slow, for whether L life or D life wins out, this autocatalysis is clearly important. This, of course, is one way to explain how a very small excess of, say, an L product might cause the reaction that produced the L product to go faster. So, for our specific case, the reaction of two L-amino acids, for example, to form the beginnings of a living system might be "catalyzed" by the little chiral excess that we started with, and it could take off and beat out the mirror-image reaction. *Voilá!*

However, do we even need an excess? A perfect example of an autocatalytic process that produces chiral substances is the formation of chiral crystals from saturated solutions of achiral or racemic molecules. One example is the crystallization of racemic of 1,1'- binaphthyl [14,15]. The R-(+) and S-(−) enantiomers are drawn in Figure 3.2.[1] For this system we have the additional "advantage" that when this molecule is melted or dissolved in a solution, the two rings are able to rotate around the bond that connects them so that they can fairly easily interconvert. We will designate this by the double arrow shown in the following chemical equation.

$$R\text{-}(-)\text{-}1, 1'\text{-binaphthyl} \rightleftharpoons S\text{-}(+)\text{-}1, 1'\text{-binaphthyl} \qquad (3.1)$$

---

[1]The CIP (R/S) system is applicable to molecules of this type, which are sometimes referred to as possessing axial chirality. The reader is referred to any good organic chemistry textbook section on nomenclature for additional information.

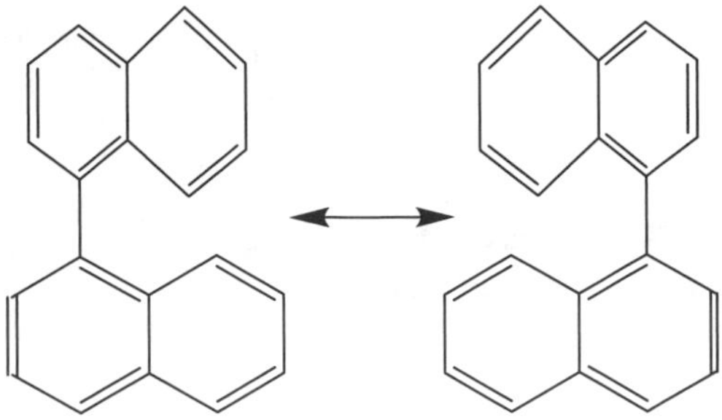

R-(−)-1,1'-Binaphthyl   S-(+)-1,1'-Binaphthyl

FIGURE 3.2. Interconverting enantiomers of 1,1'-binaphthyl.

When the solution is cooled, chiral crystals are formed that cannot interconvert. So what do you think happens when crystals are formed from the melted solution? Do you think that we always get an equal number of the two possible chiral crystals? The answer is that sometimes the crystals are almost all (+)-rotating, and sometimes almost all (−)-rotating with a statistical distribution between the two limiting situations. The average of all the crystallizations is essentially 0, as we probably should expect. This is, indeed, an autocatalytic process, because once the crystallization begins to produce (+)-crystals the equilibrium given above shifts to produce more (+)-molecules, and more (+)-crystals are formed. The real message here is that one can influence the direction of a process that potentially could produce R or S molecules. This is what most of this chapter has been about. The fact that 1,-1'-binaphthyl interconverts in solution is an important characteristic of this system, since otherwise the total amount of L- and D-crystals formed would be racemic if the initial solution were racemic. However, this isn't too restrictive or unusual. It is not difficult for a chemist to postulate autocatalytic reactions in solutions from achiral starting material that might form the basis of a self-replicating chiral "living" system.

GETTING STARTED IN THE RIGHT (OR LEFT) DIRECTION

So we can generate chiral material from racemic or achiral material quite easily. It's just that we couldn't predict ahead of time which one we might get—that is, unless you have an external chiral influence! We discussed this general idea when we described the chiral photodestruction of molecules in interstellar space. The chiral force here is the circularly polarized light that is expected to preferentially destroy one enantiomer more rapidly the other. Chemists know that if you have a saturated solution that is cooling down and ready to crystallize, you can influence which crystal you get from the two racemic possibilities by "seeding" the solution with a chiral seed crystal. This will often initiate the formation of the desired enantiomeric crystal, and then the autocatalysis takes over and more of these enantiomeric crystals are formed.

Let's go back to our pool containing the molecules that we need to get life started, and think about the possible external chiral influences. We stated earlier that perhaps our region of the universe contains a certain amount of a particular polarization (right or left) of circularly polarized UV light that could lead to an excess of L-amino acids. We presume that the universe as a whole is producing an equal amount of left and right circular polarization. How about localized chiral influences on earth? We have already mentioned that sunlight is partially circularly polarized due to the influence of the earth's magnetic field and scattering by clouds and water surfaces on the light coming from our sun. Because we have north and south poles, some parts of the earth will see a preference for right polarization and other parts, for left polarization. The amount and direction of circular polarization depends mainly on whether the pond is located in the northern or southern hemisphere, but there are other factors. Also, of course, in the earth's history the poles have changed orientation more than once. Some of the values reported for the earth's circular polarization have been as large as a few percent. Perhaps in the billions of years available for life to start, a particular pond by chance happened to start it all, and started in the direction of L-amino acids and D-sugars?

If we allow ourselves to consider that perhaps local primordial chiral influences led to a particular chirality that was a chance event, then we need to consider another, more physical, phenomenon. In order to be a "chiral effect," the force should have a preferential helical component. One possibility is circular motion (horizontal stirring!) in a vertical gravitational field. It is not difficult to imagine an inlet of a large body of water in which the waves, prevailing wind, and geometry lead to a circulation that is in predominantly one direction. These certainly exist throughout the world today! This circular motion with gravity would present a helical environment for any chemistry to take place, but, of course, over the surface of the entire planet we would expect an equal number of left and right circulating environments. Another suggestion is the Coriolis force that results from the combination of gravity and earth rotation. This force is quite small, but is helical in nature. In the southern hemisphere the direction of the force is exactly opposite, but the Coriolis force does provide a prebiotic pool with local chirality. There is also local geologic chirality. We already know that quartz can occur in chiral forms, and there are many other minerals that could serve as a chiral template for these first set of reactions that we are thinking about. Once again, the net morphology of the surfaces containing our chiral pool will average out to racemic over the entire earth, but the particular pond where life began could very well have provided a chiral template.

Summary

At this point in history the "origin of chirality" is uncertain. We really can't be sure whether our chirality is due to the fact that for billions of years chiral dust rained down on our planet, if the consequences of the non-parity-conserving weak nuclear interaction are a factor, or if we are an L-amino acid–based form of life because of a chance event. We won't have to wait too long (2014) for the Philae lander to probe the chirality of the organic molecules on the 69P/Churyumov-Gerasimenko comet. If all goes well, and all the instrumentation works, we may learn that in our region of space, L-amino acids do dominate. This result would be pretty convincing proof that the source of our chirality is extraterrestrial. If, on the other hand, the measurements show the presence of chiral organic molecules that are present in a racemic mixture, then we need to accept the possibility that the specific chirality of our biochemistry is due to a chance event. In this case we should look forward to meeting alien life forms with opposite chirality.

Suggestions For Further Reading

Popa, R., Between necessity and probability: Searching for the definition and origin of life, in *Advances in Astrobiology and Biogeophysics*, Springer-Verlag, Berlin, 2004, Chapter 5.

MacDermott, A. J., The origin of biomolecular chirality, in *Chirality and Natural and Applied Sciences*, W. J. Lough and I. W. Wainer, eds., Blackwell Science, 2002, Chapter 2.

Chela-Flores, J., Owen, T., and Raulin, F., eds., *First Steps in the Origin of Life in the Universe*, Kluwer Academic Publishers, Dordrecht, 2001.

Mason, S. F., *Molecular Optical Activity and the Chiral Discriminations*, Cambridge Univ. Press, Cambridge, UK, 1982, Chapter 11.

REFERENCES

1. Miller, S. L. and H. C. Urey, Organic compound synthesis on the primitive earth, *Science* **130**: 245 (1959).

2. Miller, S. L., Production of some organic compounds under possible primitive earth conditions, *J. Am. Chem. Soc.* **77** (9): 2351–2361 (1955).

3. Javoy, M., Where do oceans come from? *Comptes Rend. Geosci.* **337** (1–2): 139–158 (2005).

4. Cronin, J. R. and C. B. Moore, Amino acid analyses of the Murchison, Murray, and Allende carbonaceous chondrites, *Science* **172**: 1327–1329 (1971).

5. Engel, M. H. and B. Nagy, Distribution and enantiomeric composition of amino acids in hte Murchison meteorite. *Nature* **296**: 837–840 (1982).

6. Cronin, J. R. and S. Pizzarello, Enantiomeric excesses in meteoritic amino acids, *Science* **275**: 951–955 (1997).

7. Bailey, J., Astronomical sources of circularly polarized light and the origin of homochirality, *Orig. Life Evol. Biosphere* **31**: 167–183 (2001).

8. Thiemann, W. H.-P. and U. Meierhenrich, ESA Mission ROSETTA probe for chirality of cometary amino acids, *Orig. Life Evol. Biosphere* **31**: 199–210 (2001).

9. Gardner, M., *The New Ambidextrous Universe, 3rd rev. ed.*, Dover, Mineola, NY, 2005.

10. Wu, C. S., E. Ambler, R. W. Hayward, D. D. Hoppes, and R. P. Hudson, Experimental test of parity conservation in beta decay, *Phys. Rev.* **105**: 1413 (1957).

11. Lee, T. D. and C. N. Yang, Possible interference phenomena between parity doublets, *Phys. Rev.* **88**: 101 (1956).

12. Tranter, G., The effects of parity violation on molecular structure, *Chem. Phys. Lett.* **121** (4–5): 339–342 (1985).

13. Bonner, W. A., Enantioselective autocatalysis IV. Implications for parity violation effects, *Orig. Life Evol. Biosphere* **26**: 27–46 (1996).

14. Pincock, R. E., R. R. Perkins, A. S. Ma, and K. R. Wilson, Probability distribution of enantiomorphous forms in spontaneous generation of optically active substances, *Science*, **174**: 1018 (1971).

15. Pincock, R. E. and K. R. Wilson, Solid state resolution of racemic 1,1'-binaphthyl, *J. Am. Chem. Soc.* **93**: 1291 (1971).

16. Bruckner, H., P. Jaek, M. Langer, and H. Godel, Liquid chromatographic determination of D-amin acids in cheese and cow milk. Implication of

starter cultures, amino acid racemases, and rument microorganisms on formation, and nutritional considerations. *Amino Acids*, **2**: 271–284 (1992).

17. Goesmann, F., H. Rosenbauer, R. Roll, C. Szopa, F. Raulin, R. Sternberg, G. Israel, U. Meierhenrich, W. Thiemann, and G. Munoz-Caro, COSAC, The Cometary Sampling and Composition Experiment on Philae, *Space Sci. Rev.* **128**: 257–280 (2007).

LIST OF SIDEBARS AND FIGURES

Sidebar 3.A. Alice's mirror-image milk

Sidebar 3.B. The Rosetta mission

Figure 3.1. Mirror images of a spinning sphere and a spinning cone

Figure 3.2. Interconverting enantiomers of 1,1-binaphthyl

CHIRAL CHEMISTRY, RECOGNITION, AND CONTROL IN LIVING SYSTEMS

In the previous three chapters we have tried to provide a careful definition of what we mean by *chirality*, briefly discussed the scientific breakthroughs that form the basis of our current understanding of chirality in living systems, and outlined the various possible origins of chirality on earth. In the rest of this book we will try to connect our knowledge of molecular handedness to our human experiences in a chiral world. As we will soon see, there are in fact very few things in our lives that are not affected by chirality. In the remaining chapters we will present some examples to illustrate this point.

Before discussing macroscopic chirality, we should first discuss some aspects of the evolution of chirality from the very primitive self-replicating chemical system to the organized living cell and then more recognizable forms of life.

CHIRAL CHEMICAL EVOLUTION

There has been much speculation, and some theoretical and experimental research, on the incredibly complex question of the evolution of the first chiral L-amino acids, D-sugars, or precursors to these molecules, to the first true forms of "life." In Chapter 3 we presented a number of different scenarios for getting chiral starting material, but the chemistry, physics, and biology of putting these building blocks together with an energy source to organize, differentiate, self-replicate, and perform other functions is far from being understood. One can find interesting research papers and book chapters on very complex chemical paths, statistical arguments, and so on to try to come up with a plausible mechanism for the development of true "living systems," but this is not the main topic of this book, and we refer the reader interested in exploring this question in more detail to some suggested readings at the end of this chapter.

Is it surprising that we know so little about this chiral chemical evolution? Not really. Modern chemists' understanding of chemical reactions, including many important biological reactions, is getting better all the time, but chemists work in a finite lifetime, so reactions that are not very likely to happen are not the kinds of systems that can be studied or examined in much detail. The chemical reactions that led to the formation of L-amino acids and D-sugars, and eventually to the helical proteins and RNA and DNA involved billions of years, a sample size as large as the volume of water on earth, and energy from sunlight, temperature, lightning, cosmic radiation, and other sources on which we can only speculate. Since, of course, we know where all of this chemistry ended up (life as we know it), we do know some aspects about the overall reactions that were successful in leading to life. This knowledge comes from the determination of the structures that we find in living systems, such as the L-amino acids, D-sugars, and right-handed helices; the larger units of nuclei, cells, and organs; and the final myriad of "living" units, including plants, animals, and humans. The more sophisticated our understanding of biochemistry and molecular biology becomes, the more scientists will be able to develop reasonable models for the beginning of life, but this will be just speculation for many years to come.

It should already be obvious that the initial "selection" of chiral starting material whether it was L-amino acids, D-sugars, or their chiral precursors, set life on a path in which chiral molecular structures play a critical role. The biochemistry of life requires an ability to recognize specific molecular structures (most of which are chiral), and then to initiate specific biochemical responses. In this chapter we will describe some general properties of chiral recognition, and then discuss the importance of molecular chirality on smell, taste, and other aspects of chiral molecular interactions.

CHIRAL RECOGNITION AT THE MOLECULAR LEVEL

So far we have presented and discussed only static aspects of molecular structure. Hopefully by now the reader can identify (with help, perhaps) when objects such as molecules are chiral. From the beginning of this book we have argued that the complexity of life requires that the molecules of life, for example amino acids, be exclusively one chirality. This principle is sometimes referred to as *homochirality*. Of course, this doesn't mean that every chiral center must be R and not S, or $(+)$ and not $(-)$, but it does mean that the chirality of the molecular substituents of life matters. The general reasoning here is that the important biochemical reactions that allow living systems to survive, propagate, react to surroundings, and so on must be under precise molecular control, and this could be possible only if the full three-dimensional structures of the reacting substances are involved and exploited in the necessary chemical reactions.

To understand and test this principle, chemists have constructed simple models to describe the interactions of chiral

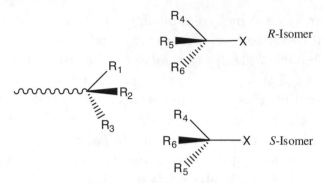

FIGURE 4.1. Diastereomeric interactions.

molecules. We present here a simplified version of the so-called three-point model to help in developing an appreciation of the principle of chiral recognition. Consider the interaction of the molecules drawn in Figure 4.1. We are again using the chemical drawing convention that solid lines are in the plane of the paper, black wedges are used to show connections that extend out of the plane of the paper, and the hashed lines are meant to illustrate a molecular bond that projects away from the reader. The drawing on the left is intended to represent a tetrahedral carbon attached to a larger molecular fragment, for example, a protein, and the two molecules are mirror-image enantiomers. Without specifying in any detail what happens when one of the molecules on the right approaches the larger species on the left, it should be evident that the "interactions" in these two situations will be different. We assume that the molecule on the right is free to rotate about the "C–X bond", so the possible interactions of the two species are as listed in Table 4.1.

The types of interactions that could be important in these situations are electrostatic (i.e., the repulsion of similar electric charges and attraction of opposite charges), weak attractive forces like the "hydrogen" bonding that we have seen in the structures of proteins and DNA, or the formation of strong or

TABLE 4.1. POSSIBLE PAIRWISE INTERACTIONS OF MOLECULES
PICTURED IN FIGURE 4.1

R-isomer	S-isomer
$R_1 : R_4 + R_2 : R_5 + R_3 : R_6$	$R_1 : R_4 + R_2 : R_6 + R_3 : R_5$
or	or
$R_1 : R_5 + R_2 : R_6 + R_3 : R_4$	$R_1 : R_5 + R_2 : R_4 + R_3 : R_6$
or	or
$R_1 : R_6 + R_2 : R_4 + R_3 : R_5$	$R_1 : R_6 + R_2 : R_5 + R_3 : R_4$

weak "covalent" bonding in which electrons are shared between the interacting atoms, or they might be simple structural
considerations due to the size and shape of the approaching
species. This last aspect is usually referred to by chemists as
steric factors.

Let's make our model slightly more specific by referring to
Figure 4.2. We now arbitrarily assign charges to two of the three

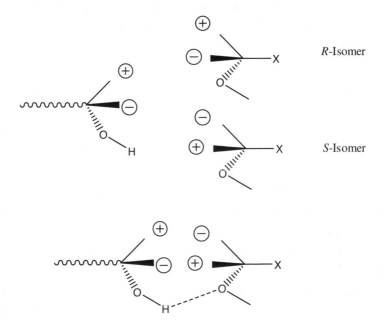

FIGURE 4.2. Example of a "lock and key" diastereomeric interaction.

R-groups on each molecule, and also illustrate a common type of "hydrogen bonding" in which the hydrogen atom of a hydroxyl group (OH) is weakly attracted to another oxygen atom. Looking at the all the possible pairwise interactions as listed in Table 4.1, we can see that the preferred (most stable, lowest energy, etc.) arrangement is the S-isomer in the orientation as shown in the bottom of Figure 4.2. When two species such as those shown in Figures 4.1 and 4.2 interact, many things can happen. For example, rearrangements can occur and chemical bonds can be formed or broken. These types of chemical recognition/responses are central to the biochemistry of life. We have described this simple model only to illustrate the main point that chiral substances may react differently when exposed to or contacted by other chiral substances.

The general idea that biochemical processes follow specific chemical recognition/response events was first proposed by Emil Fischer, who used the term "lock and key." The implication here is that a specific chemical compound "fits" into a specific location in a biochemical and that this then initiates a certain biochemical response. The term "lock and key" is still widely used today, but there is a much better appreciation nowadays that there are a range of responses to chemicals, and that many different molecules with similar, but not necessarily identical, structures can initiate similar biochemical responses. However, the chirality of the interacting species is important in almost all examples of this principle.

In the examples that follow in this chapter, we seldom know the precise molecular mechanism that leads to the discrimination between enantiomers; we only see the result of this differential diastereomeric interaction. As we shall see, the study of these interactions is a powerful way to develop new medicines, flavors, and other important biologically active materials.

EXAMPLES OF CHIRAL RECOGNITION AND DISCRIMINATION

SMELL

Much progress has been made by scientists in developing an understanding of the physical chemical mechanisms involved in the human sense of smell [1,2]. It is now believed that when volatile compounds enter the nose they dissolve in an aqueous/lipid environment of mucus on the roof of each nostril on an olfactory epithelium region that is only about 5 cm^2 (0.75 in.^{2}) in humans. Within this mucus the molecules become bound to odorant binding proteins (OBPs), which transport them through the mucus layer and bind to odorant receptors (ORs). It is estimated that in humans there are approximately 1000 different types of ORs, and the 5-cm^2 olfactory sensing region of each nostril contains more than 10 million ORs. More recent research has shown that each type of receptor is sensitive to more than one structural type, and each odorant activates more than one receptor. So scientists believe that the sense of smell involves a very complicated code that is forwarded to the brain from the olfactory bulb, which is interpreting all the signals received from the ORs.

The odorant receptor proteins and the odorant receptors themselves are composed of L-amino acids, so it should not be surprising that the nose can be quite sensitive to the chirality of the odorant molecules, in addition to many other aspects of size, shape, and chemical structure. Louis Pasteur was actually the first person to suggest that the olfactory senses might be dependent on the dissymmetry of the molecule involved [3]. Progress in understanding the chiral selectivity of smell has been difficult because of the importance of studying pure compounds. In fact, even in the 1960s a theory of olfaction was developed by Wright, who proposed the importance of

molecular vibrations in the far-infrared region of the electromagnetic spectrum in odor detection [4]. Part of his reasoning
to ascribe odor detection to a physical rather than chemical
phenomenon was the lack of any confirmed examples of chiral
selectivity in odor detection. The vibrational frequencies of
enantiomers are identical.

As we saw with the early measurements of Biot, natural
sources of chiral material such as lemon oil are complex mixtures. We will see shortly that not only are there differences in
the smell of many enantiomers; there are also large differences
in threshold sensitivity, that is, the concentration in which the
substance can be detected by the nose. So you might have a
99.9% pure sample, but the dominant aroma may be from the
0.1% impurity. It wasn't until the midtwentieth century that
chemists were able to prepare "pure" compounds of enantiomers and then verify the different smells. One of the first
examples of such compounds are R-(−)-carvone and S-(−)-
carvone, which were studied by three different research teams
and published independently in 1971 [5–7]. The structures of
these enantiomers are shown in Figure 4.3. It is important to
remember that these two compounds have identical physical
properties unless they are being analyzed or reacted with

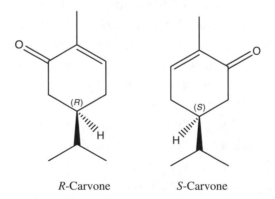

R-Carvone   S-Carvone

FIGURE 4.3. Molecular structures of R- and S-carvone.

something that is chiral. However, R-(−)-carvone is the smell you experience from spearmint leaves, and S-(−)-carvone is what you smell in caraway seeds. There are now many known examples of this difference in the smell of enantiomers. We will mention only a few in this chapter. Dr. John C. Leffingwell has collected odor sensitivity data and descriptive smells on more than 925 chiral compounds on a freely accessible Website [8]. Another example of chiral differences in smell is the molecule R-(+)-limonene introduced in Chapter 2 which has an orange aroma, whereas S-(−)-limonene smells like lemon. Sometimes the difference between enantiomers is more in the sensitivity than the smell. The compound (+)-nootkatone, for example, drawn in Figure 4.4 is responsible for the odor of grapefruit. The sensitivity threshold for smelling this compound is approximately 2000 times lower than its enantiomer, which also has a fruity aroma [9].

Another interesting example is the compound androstenone. As indicated in Figure 4.5, this is a large molecule containing six chiral carbon centers. This molecule is almost the largest capable of being smelled by a human. As the name implies, the (+)-enantiomer is produced by male humans, and this compound has been marketed in a spray form as a human female sex attractant, although the smell is described by many people as having the characteristic odor of urine. The compound certainly attracts pigs; in fact, pigs are able to find truffles

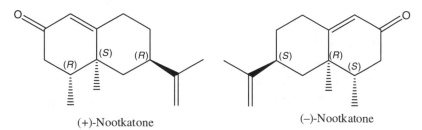

(+)-Nootkatone   (−)-Nootkatone

FIGURE 4.4. Molecular structures of (+)- and (−)-nootkatone.

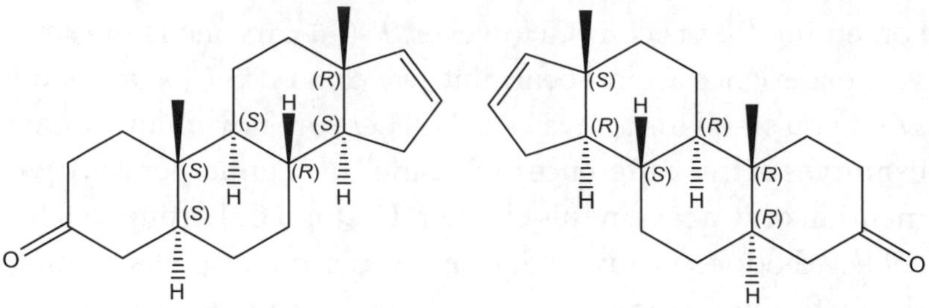

(+)-Androstenone   (−)-Androstenone

FIGURE 4.5. Molecular structures of (+)- and (−)-androstenone.

through the (+)-androstenone that truffles emit. The (−)-enantiomer apparently has no detectable odor for humans. Complicating the use of this compound as a sexual attractant is the fact that close to 50% of us can't smell either enantiomer! The inability to smell certain compounds or the variability in our description of the smell that we perceive makes the studies of smell quite difficult. Tests are usually performed on panels of people, and the results are summarized or averaged.

Our noses can obviously be very sensitive detectors of chirality, but there are few examples as obvious as the enantiomers of carvone. Perusal of the Leffingwell database illustrates this point as, the terms used to describe the different enantiomeric smells are often very similar. It has been shown that animal species other than humans can demonstrate chiral selectivity better than humans can. This is to be expected because of the greater importance of this sense to these animals. Trained human sniffers are better able to discriminate between subtle differences in smell than most of us are, and it is not surprising that much of the current research in this area is associated with the perfume industry. Some of the impetus for obtaining more information concerning the structure–smell relationships is, of course, driven by economic considerations and market competition among perfume producers. However, there is also increasing awareness and concern for the environment, and detailed knowledge of the effects of all ingredients in perfumes, including properties of individual enantiomers, is obviously important. An excellent review emphasizing the enantiomeric differences in smell appeared in 2006 [10].

Taste

The sense of taste is in some respects simpler than smell. It is generally believed that we can detect only five different types of taste, namely salty, sour, sweet, bitter, and umami. The fifth taste type, umami, is relatively new to Western knowledge of taste, but it has been known in Japan from the early 1900s. It is variously described as "meaty," savory," or like "aged cheese", and is associated mostly with monosodium glutamate, (MSG) which was discovered and isolated by Kikunae Ikeda from seaweed in 1908 [11]. Umami has been elevated to one of the taste types following the identification of umami taste receptors in 2002 at the University of Miami [12]. The senses of taste and smell are strongly connected, and in scientific experiments aimed at exploring the sensitivity of taste receptors, subjects are usually provided with some type of nose clamps to reduce the influence of odor. Just as in the sense of smell, the perception of taste depends on who is doing the tasting. There are much fewer data available on the influence of chirality on taste than on the influence of chirality on odor. Perhaps it is less interesting to humans to voluntarily place unknown substances on their tongues, than to sniff them in the air. It is perhaps not too surprising to hear that the artificial sweeteners saccharin, aspartame (NutraSweet), and cyclamate (now banned by the FDA) were all discovered by accidentally getting these compounds into the mouth through bad laboratory practices.

It should not be a surprise to learn that taste can also be sensitive to chirality. A good example is, in fact, MSG. This compound, which may be purchased at the supermarket and

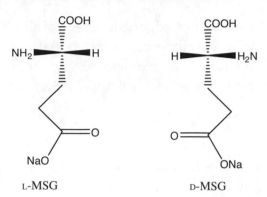

L-MSG   D-MSG

FIGURE 4.6. Molecular structures of monosodium glutamate (MSG) enantiomers.

added to food, is actually the mono- (i.e., one-) sodium salt of the amino acid L-glutamic acid shown in Figure 4.6. Mono-sodium-D-glutamate has no taste. In a relatively recent publication on the chiral selectivity of taste, Shallenberger states that there is some selectivity on the taste of amino acids, but that both D- and L-sugars taste sweet [13]. In a more recent taste test by "trained" subjects of 18 of the 20 essential amino acids, it was reported that some D-amino acids had more taste and some less total taste than did the naturally occurring L-isomer [14]. There was no obvious correlation between chirality and sweetness, sourness, or bitterness. Many of the L-amino acids displayed strong umami, but none of the D-amino acids displayed any umami taste.

Some years ago, it was suggested that a good "artificial" sweetener would be the enantiomer of table sugar, D-sucrose, which is composed of D-fructose and D-glucose. It would taste sweet, but since it is the opposite enantiomer, it would not be fattening since it would not be digested and metabolized in the same way as natural sugar. This is a reasonable idea, since both D- and L-sugars are known to be sweet-tasting. The problem is that the manufacturing of a nonnatural sugar on the scale necessary to be commercially viable is nontrivial, and the costs are currently prohibitive. No doubt there are some entrepreneurial synthetic chemists out there looking for cheap ways to make L-sugars.

There is much to be learned about the chiral selectivity involved in taste. As was described for the sense of smell, our perception of how something tastes might be a coded response from the five types of taste receptors. There is also much to be learned about the way in which specific chemical structures and types (acids, alcohols, etc.) are recognized in the taste buds. Research in this area is, of course, being led by agricultural and food chemists. They are beginning the kind of systematic chemistry that is needed to probe the structure of taste receptors and the mechanism of taste perception. The importance of chirality has been recognized by agricultural and food chemists, and it is now expected that research publications in this area will include identification and physical properties of individual enantiomers. Increasingly, scientists are using animal models to probe the reaction of taste receptors to enantiomers and other compounds. In these experiments, of course, the fish, for example, are not marking tasting scoring cards, but are connected to electric devices to measure nerve stimuli.

COMMUNICATION

We humans use our senses of smell and taste very differently than do other species. Compared to most other lower animals, the sensitivity and selectivity of our noses is quite poor. The area of the olfactory region for cats, for example, is 5 times larger that that for humans. We, of course, rely primarily on our eyes and ears to avoid danger, use our sense of taste to avoid swallowing poisons, and have the ability to communicate threats or needs through a well-developed language. We will avoid places or things that have unpleasant odors, and we enjoy the pleasant aromas of flowers and good food, but we rely on smell mostly in situations where we can't use hearing or sight. Many species use the production and the detection of odors for communication, and the importance of chirality in these communication systems is widely recognized.

When an individual animal or plant secretes a substance that is detected by an individual of the same species and induces a specific reaction, the substance is said to be a *pheromone* [a term derived from the Greek *pherein* (to transfer) and *hormon* (to excite)]. The first report of such chemical communication was by the French naturalist Jean-Henri Fabre (Biographic Photo 4.1), who observed the attraction of male great peacock or emperor moths to a caged female (see Sidebar 4.A). Pheromones have been associated mainly with various aspects of sexual attraction and reproduction, but also

BIOGRAPHIC PHOTO 4.1. Jean-Henri Fabre. (Public-domain photograph obtained form the Library of Congress.)

have been identified in alarm signaling, aggregation, and other behaviors. These kinds of chemicals are also observed in individual identification, and in attraction of prey or repulsion of predators. Communication in an efficient way by means of volatile compounds to send a specific message, for a specific species, and sometimes for a specific gender is a complex problem. The two main approaches are for the communicating species to compose the intended message by varying the relative concentrations of a small number of volatile compounds, or to emit or secrete a specific compound often with a specific enantiomeric identity. Although pheromones have been studied in mammals, especially rodents and pigs, the largest amount of research, especially work concerned with the importance of chirality, has involved insects. We will describe some of this work, since it illustrates the many ways in which chirality affects chemical communication.

Sidebar 4.A

The Night of the Great Peacock. Jean-Henri Fabre was a self-educated naturalist who is best known for being one of the pioneers in the study of insect behavior. One May morning in the 1870s he watched a female great peacock moth emerge from her cocoon. He trapped the moth in a wire-gauze bell jar, as was his habit in order to observe the moth spread her wings, which were almost 6 inches across. That night around nine o'clock, his young half-dressed son "Little Paul" starting jumping and stomping around, yelling for his father to come and see some moths that were as "big as birds." More than 150 male moths showed up over the next few nights in pursuit of the female. Through careful observation and experimentation, Fabre was able to conclude that the male moths were not being attracted by light, either visible or nonvisible, nor by a sound because of the long distances that the moths traveled. He surmised that it must be a smell emanating from the female. This is the first experimental report of an insect pheromone, and the correct explanation of what he called "the night of the great peacock" [24].

Great peacock moth. (Illustration by Rachel MaKarrall.)

Professor Kenji Mori of the Science University of Tokyo has been the leading researcher in the study of chirality in insect pheromones [15]. While teaching Pasteur's very first experiments on stereochemistry to undergraduate organic chemistry students in 1973, Professor Mori realized that he could use the two forms of tartaric acid to synthesize the two enantiomers of the compound brevicomin that are drawn in Figure 4.7. In collaboration with colleagues at the University of California (UC) Berkeley, he was amazed to see how attractive (+)-brevicomin was to western pine beetles, and how uninterested

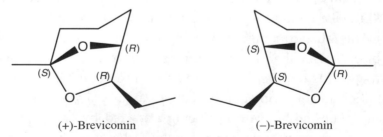

(+)-Brevicomin   (−)-Brevicomin

FIGURE 4.7. Molecular structures of (+)- and (−)-brevicomin.

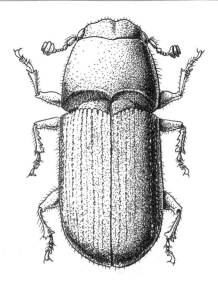

Western pine beetle. (Illustration by Rachel MaKarrall.)

these beetles were in (−)-brevicomin [16]. These initial experiments led to many others in Professor Mori's lab and others on discovering the various ways in which insects exploit chirality in their communication systems.

Critical to all experiments concerned with naturally occurring pheromones and related compounds is the identification of the compounds, especially the chirality of the various chiral centers. Starting with the very first determination of the absolute structure of (+)-tartaric acid, single-crystal X-ray diffraction has been used extensively to determine absolute chemical structure. This technique is available, however, only if the substance can be obtained in pure form, and in sufficient quantities to grow single crystals that are large enough to make this measurement. Actually, even if you have a large amount of a pure compound, sometimes it just does not form crystals suitable for application of this technique. For the study of pheromones in insects, the quantity of material available is usually very small (ranging from micrograms to a few milligrams), and the compounds themselves are volatile oils that do not readily crystallize. There are ways to probe chiral structure using spectroscopic techniques, but unfortunately these seldom allow one to determine unambiguous chiral structures. If you are unable to grow suitable crystals for structural analysis by X-ray diffraction, then the only way to ensure that you have the structure correct is to synthesize possible compounds starting with structures of known chirality, making sure that you know the *R* or *S* identity of chiral centers throughout the synthesis, and then compare the final optical rotation and other properties with the natural pheromone.

It seems as though insect communication systems have exploited chirality in every way possible. We have already seen the simplest case where one enantiomer, (+)-brevicomin, is attractive to the western pine beetle and the other enantiomer, (−)-brevicomin, is not. It has also been observed that males of the scratch-faced ambrosia beetle *Gnathotrichus sulcatus* of the Pacific coast of North America produce both enantiomers of the aggregation pheromone sulcatol drawn in Figure 4.8 [17]. It

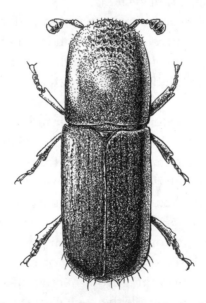

Scratch-faced ambrosia beetle. (Illustration by Rachel MaKarrall.)

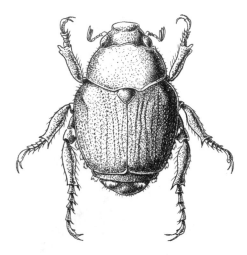

S-Sulcatol

R-Sulcatol

FIGURE 4.8. Molecular structures of R- and S-sulcatol.

Japanese beetle. (Illustration by Rachel MaKarrall.)

turns out that neither enantiomer is active on its own, but that the aggregation requires a mixture of the two. There are other examples of this synergistic effect. In the Japanese beetle (*Popillia japonica*) the active sex pheromone is R-japonilure (Figure 4.9), but the racemic mixture is inactive [18]. Apparently S-japonilure inhibits the effect of its enantiomer. These two examples illustrate how insects can use the extent of chirality as a control mechanism.

Another interesting situation is the olive fruit fly (*Bactrocera oleae*). The female produces the sex pheromones R-olean and S-olean as drawn in Figure 4.10 [19]. The R-olean excites males, and the S-olean excites herself! One final example of the amazing variety of ways in which insects use chirality is the

R-Japonilure

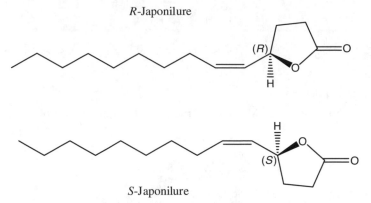

S-Japonilure

FIGURE 4.9. Molecular structures of R- and S-japonilure.

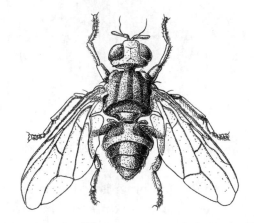

Olive fruit fly. (Illustration by Rachel MaKarrall.)

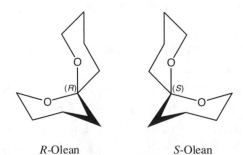

R-Olean   S-Olean

FIGURE 4.10. Molecular structures of R- and S-olean.

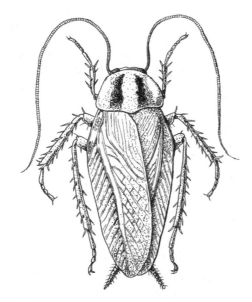

German cockroach. (Illustration by Rachel MaKarrall.)

FIGURE 4.11. The *S,S*-isomer of the *Blatella germanica* pheromone.

German cockroach (*Blattella germanica*). The female produces the contact pheromone shown in Figure 4.11 with the specific chirality at the two carbon centers shown [20]. Note that this is a very big molecule (the subscript 17 denotes a 17-carbon chain) and is not volatile, so it must be touched (tasted?) to be effective. The male German cockroach, it seems, doesn't care about chirality, as he is attracted to all four possible isomers *RR*, *RS*, *SR*, and *SS*. Further amazing insect stories await the reader of the additional references listed at the end of this chapter.

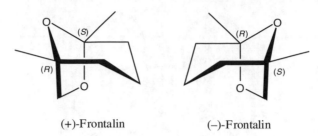

(+)-Frontalin   (−)-Frontalin

FIGURE 4.12. Molecular structures of (+)- and (−)-frontalin.

Of course, the importance of pheromone chirality is not limited to insects. Asian bull elephants secrete a large quantity of chemicals from a temporal gland on the face during periodic episodes of heightened sexual activity and aggression known as *musth*. It has been noted that the ratio of the enantiomers of the pheromone frontalin (Figure 4.12) varies during the elephant's maturation from adolescence to adulthood [21]. Young bulls secrete more (+)-frontalin than (−)-frontalin, and adult bulls secrete almost equal amounts. The racemic secretions were shown to be attractive to fertile female elephants, whereas the (+)-frontalin-enriched secretions were not.

Knowledge about the biology and chemistry associated with chemical communication is becoming increasingly important as society tries to develop more environmentally sensitive methods of pest control. For example, knowledge of the specific sensitivities of mosquitoes and other insects to specific enantiomers of repellents will no doubt lead to more targeted commercial products [22]. Certainly the use of pheromones in large-scale commercial operations is increasing, due to efforts to increase food production, while decreasing the widespread use of pesticides and herbicides. For example, the catalog of Great Lakes IPM (insect pest monitoring) Inc.[1] sells pheromone insect traps for 66 different insect species. In many cases the

---

[1]Further information is available online at www.greatlakesipm.com.

pheromones are used to attract specific insects as a way of determining whether insecticides need to be sprayed, but in other commercial applications pheromones are used to directly affect insect populations by disrupting mating behavior through widespread application of pheromones that make it very difficult for mating partners to locate each other.

In most large-scale commercial applications of phero-mones, the active pheromone is not chiral, a synthetic achiral pheromone of a similar structure is active enough, or a racemic mixture is used. The synthesis and separation of enantiomers is usually too costly for large-scale use, and if the racemic mixture or achiral substitute is effective, this presents a much more economically feasible pest control method (see Sidebar 4.B).

Sidebar 4.B

Gypsy Moth Traps. The gypsy moth first appeared in the United States in 1869, and by 2008 had infested most of the East Coast and Wisconsin, and is also becoming a concern in Minnesota. In the caterpillar stage, the insect eats leaves from a wide variety of trees, leaving areas of widespread defoliation. Many trees die because of the resultant stress. In 2007 the State of Minnesota set more than 15,000 gypsy moth traps containing gypsy moth pheromone in order to monitor the spread of the insect. The enantiomer (+)-disparlure shown below is the enantiomer that is the sexual attractant to male gypsy moths. No effect on male gypsy moth behavior is seen with the enantiomeric compound (−)-disparlure, and because of the much more economical synthesis of the racemic mixture, this is what is contained in the commercial gypsy moth traps. If the traps detect a significant increase in population, then a large application of disparlure is used to disrupt mating behavior.

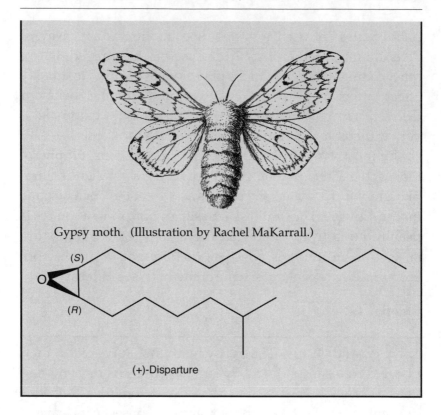

Gypsy moth. (Illustration by Rachel MaKarrall.)

(+)-Disparture

So far we have completely avoided the issue of whether there are human pheromones. The discussion given earlier in this chapter about androstenone, of course, suggests that at least some people believe that they do exist. Certainly there is no evidence for the level or complexity of chiral selectivity in chemical communication in humans that is widespread in insects. In dating (or mating) situations, humans, of course, rely on visual cues and voice communication.

CHIRALITY AS A CONTROL MECHANISM IN BIOCHEMISTRY

Once we step back from the external secretions or external sensors of organisms, into the intricate biochemistry of living systems, we, of course, see the importance and influence of chirality everywhere. This is what we would expect for a complex living system in which just about every important constituent is present as only one of the two possible mirror-image forms. This is not a biochemistry textbook, so we will not discuss this very much. However, living things use chirality in numerous interesting and, perhaps, unexpected ways, and we will end this chapter by describing one recent discovery involving fireflies.

Many living species give off light, but the one that we are probably most familiar with is the firefly. The chemical reaction that produces light from the firefly, so-called bioluminescence, is fairly well understood. It is known from many experiments that the molecule that is involved in the bioluminescence of fireflies is D-luciferin, which is drawn in Figure 4.13 along with its enantiomer L-luciferin. This is an old but very descriptive name for this compound. "Lucifer" is a reference to fire (and the

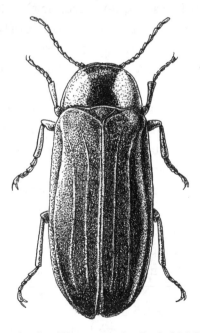

Japanese firefly. (Illustration by Rachel MaKarrall.)

D-Luciferin

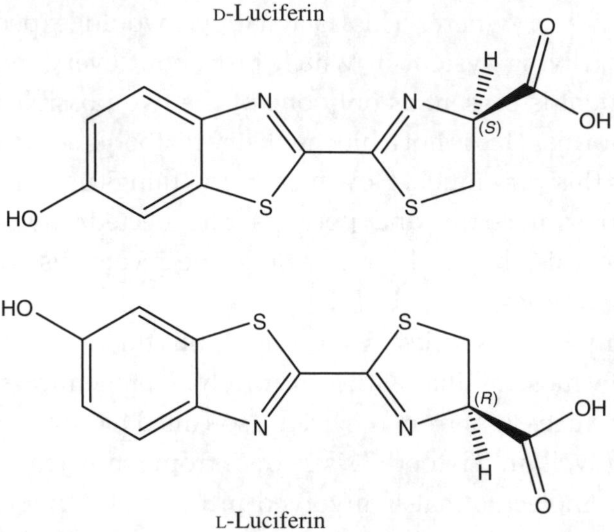

HO

HO

L-Luciferin

FIGURE 4.13. Molecular structures of D- and L-luciferin.

devil!), and D or L is used to describe the chirality of the amino acid cysteine, which is the presumed source of the chiral carbon. The structure of the naturally occurring L-cysteine and its enantiomer D-cysteine are drawn in Figure 4.14. In this figure we have drawn the amino acids as a Fischer projection, and as a three-dimensional structure. The bottom drawing has been oriented so that it is aligned with the drawing of luciferin in Figure 4.13. Hopefully you can see that D-luciferin has the designation D because of the chirality of the amino acid D-cysteine from which it was presumably made. The interesting problem here is that we know that the naturally occurring form of the amino acid cysteine is L-cysteine. So, how did the firefly make D-luciferin (or D-cysteine), and why would this phenomenon use the nonnatural isomer? These were problems investigated by Dr. Kazuki Niwa and his collaborators in Osaka and Tokyo [23]. These researchers raised Japanese fireflies (*Luciola lateralis*) and analyzed the amount of D-luciferin,

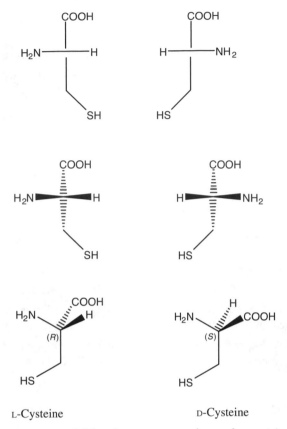

L-Cysteine   D-Cysteine

FIGURE 4.14. Molecular structures of D- and L-cysteine.

L-luciferin, D-cysteine, and L-cysteine at various stages in the life cycle of this insect. The amount of D-luciferin increases with the age of the firefly, and very little D-cysteine was detected. They concluded that the firefly synthesizes L-luciferin from the naturally occurring amino acid, and then using an available enzyme inverts the chiral carbon to make D-luciferin, which is used for the blinking action to attract a mate. Storing the necessary luciferin as the D-isomer makes it less susceptible to any other natural metabolic processes that might use the L-cysteine. This firefly does not eat after the larva stage, so the storage of the *important* luminescent compounds is necessary.

SUMMARY

It should be clear that an appreciation of the importance of chiral molecular structure and the discrimination and control provided by the interactions of chiral molecules is now essential for anyone interested in understanding insect pheromones, smell, taste, and, of course, biochemistry. In the next chapter we will discuss the critical importance of chirality in an area that impacts our lives in ways more direct than insect communication, namely, pharmaceuticals.

SUGGESTIONS FOR FURTHER READING

Wilson, D. A., and R. J. Stevenson, *Learning to Smell: Olfactory Perception from Neurobiology to Behavior*, Johns Hopkins Univ. Press, Baltimore, 2006.

Agosta, W. C., *Chemical Communication: The Language of Pheromones*, Scientific American Library, 1992.

Wyatt, T. D., *Pheromones and Animal Behaviour: Communication by Smell and Taste*, Cambridge Univ. Press, Cambridge, UK, 2003.

REFERENCES

1. Malnic, B., J. Hirono, T. Sato, and L. B. Buck, Combinatorial receptor codes for odors, *Cell* **96**:713–723 (1999).

2. Zozulya, S., F. Echeverri, and T. Nguyen, The human olfactory receptor repertoire, *Genome Biol.* **2**(6) (2001).

3. Pasteur, L., Memoire sur la fermentation appelee lactique, *Comptes Rend. Acad. Sci. Paris* **46**:615 (1858).

4. Wright, R. H., Odour of optical isomers, *Nature* **198**:782.

5. Russell, G. F. and J. I. Hills, Odor differences between enantiomeric isomers, *Science* **172**:1043 (1971).

6. Leiteregg, T. J., D. G. Guadagni, J. Harris, T. R. Mon, et al., *Nature* **19**:785 (1971).

7. Friedman, L. and J. G. Miller, Odor incongruity and chirality, *Science* **172**:1044 (1971).

8. Leffingwell, J. C., *Chirality and Odour Perception*, www.leffingwell.com/chirality/chirality.htm 2007.

9. Haring, H. G., F. Rijkens, H. Boelens, and A. J. van der Gen, Olfactory studies on enantiomeric eremophilane sesquiterpenoids, *Agric. Food Chem.* **20**:1018 (1972).

10. Bentley, R., The nose as a steroechemist. Enantiomers and odor, *Chem. Rev.* **106**:4099–4112 (2006).

11. Ikeda, K., New seasonings, *J. Chem. Soc. Tokyo* **30**:820–836 (1909).

12. Nelson, G., J. Chandrashekar, M. A. Hoon, L. Feng, G. Zhao, N. J. P. Ryba, and C. S. Zuker, An amino-acid taste receptor, *Nature* **416**(6877): 199–202 (2002).

13. Shallenberger, R. S., Taste recognition chemistry, *Pure Appl. Chem.* **68**(4): 659–666 (1997).

14. Kawai, M. and Y. Hayakawa, Complex taste—taste of D-amino acids, *Chem. Senses* **30**(Suppl. 1):240–241 (2005).

15. Mori, K., Chirality in the natural world: chemical communication, in *Chirality in Natural and Applied Science*, W. J. Lough and I. W. Wainer,eds., Blackwell Science, Ltd., Bodwin, Cornwall, UK, 2002, pp. 241–259.

16. Wood, D. L., L. E. Browne, B. Ewing, K. Lindahl, W. D. Bedard, P. E. Tilden, K. Mori, G. B. Pitman, and P. R. Hughes, Western pine beetle: Specificity among enantiomers of male and female components of an attractant pheromone, *Science* **192**:896–898 (1976).

17. Byrne, K. W., A. Swigar, R. M. Silverstein, J. H. Borden, and E. Stokkink, Sulcatol: Population aggregation pheromone in *Gnathotrichus sulcatus*, *J. Insect Physiol.* **20**:1895–1900 (1974).

18. Tumlinson, J. H., M. G. Klein, R. E. Doolittle, T. L. Ladd, and A. T. Proveaux, Identification of the female Japanese beetle sex pheromone: Inhibition of male response by an enantiomer, *Science* **197**:789–792 (1977).

19. Haniotakis, G., W. Franske, K. Mori, H. Redlich, and V. Schurig, Sex-specific activity of (R)-(−)- and (S)-(+)-1,7-dioxaspiro[5.5]unde-cane, the major pheromone of Dacus oleae, *J. Chem. Ecol.* **12**:1559–1568 (1986).

20. Mori, K., S. Masuda, and T. Suguro, Stereochemical synthesis of all of the possible stereoisomers of 3,11-dimethylnonacosan-2-one and 29-hy-droxy-3,11-dimethylnonacosan-2-one, the female sex pheromone of the German cockroach, *Tetrahedron* **41**:3663–3672 (1981).

21. Greenwood, D. R., D. Comesky, M. B. Hunt, and L. E. L. Rasmussen, Chemical communication: Chirality in elephant pheromones, *Nature* **438**:1097–1098 (2005).

22. Klun, J. A., W. F. Schmidt, and M. Debboun, Stereochemical effects in an insect repellent, *J. Med. Entymol.* **38**(6):809–812 (2001).

23. Niwa, K., M. Nakamura, and Y. Ohmiya, Stereoisomeric bio-inversion key to biosynthesis of firefly D-luciferin, *FEBS Lett.* **580**:5283–5287 (2006).

24. Fabre, J.-H., The great peacock or emperor moth, in *Fabulous Insects*, C. Neider, ed., Harper, New York, 1954.

LIST OF BIOGRAPHIC PHOTOGRAPHS, SIDEBARS, AND FIGURES

Biographic Photo 4.1. Jean-Henri Fabre

Sidebar 4.A. The night of the great peacock

Sidebar 4.B. Gypsy moth traps

Figure 4.1. Diastereomeric interactions

Figure 4.2. Example of a "lock and key" diastereomeric interaction

Figure 4.3. Molecular structures of R- and S-carvone

Figure 4.4. Molecular structures of (+)- and (−)-nootkatone

Figure 4.5. Molecular structures of (+)- and (−)-androstenone

Figure 4.6. Molecular structures of monosodium glutamate (MSG) enantiomers

Figure 4.7. Molecular structures of (+)- and (−)-brevicomin

Figure 4.8. Molecular structures of R- and S-sulcatol

Figure 4.9. Molecular structures of R- and S-japonilure

Figure 4.10. Molecular structures of R- and S-olean

Figure 4.11. The S,S- isomer of the *Blatella germanica* pheromone

Figure 4.12. Molecular structures of (+)- and (−)-frontalin

Figure 4.13. Molecular structures of D- and L-luciferin

Figure 4.14. Molecular structures of D- and L-cysteine

CHIRALITY AND DRUGS

The importance of chirality in the design, development, and use of drugs for the treatment of disease and other health-related issues cannot be overstated. Worldwide, the economic value of pharmaceuticals is on the order of trillions of dollars with individual drugs selling for multiple billions of dollars per year. Aside from the financial aspects, of course, the impact of modern drug development and commercialization thus far on the length and quality of life for humans on earth has been enormous. Although there are some very important achiral drugs, such as aspirin and Tylenol®, most pharmaceuticals have chiral structures and are available as either racemic mixtures or pure enantiomers. For the reasons outlined below, there has been a sharp increase in the approval of chiral drugs versus racemic mixtures since the mid-1990s [1]. In 2006, for example, more than 80% of new drugs approved by the FDA were chiral, and more than 75% of these were being produced and marketed as a single enantiomer [2].

In this chapter, we will first present some of the historical and regulatory aspects of chirality in drug development, present a few examples illustrating the importance of knowing

Mirror-Image Asymmetry: An Introduction to the Origin and Consequences of Chirality
by James P. Riehl
Copyright © 2010 John Wiley & Sons, Inc.

and controlling chirality, and try to make sense of a very confusing chiral drug industry. We will end with a brief discussion of the various ways that these chiral materials are manufactured.

CHIRALITY IN NATURAL REMEDIES AND FOLK MEDICINES

Natural or folk medicines are almost exclusively obtained from living plants or animals, and therefore they are almost all chiral substances that are composed of only one enantiomer. The first written report of the use of natural medicines is attributed to the Chinese emperor and scientist Shen-Nung [3]. Tradition has it that he personally tasted or ingested a large number of natural substances to determine whether there was any beneficial effect. In his treatise on herbs, which dates from 2735 BC, Shen-Nung described the medicinal use of many substances derived from plants including the treatment of fevers using the powdered root of the plant Ch'ang Shan (*Dichroa febrifuga* Lour) and the use of the leaves of the plant Ma Huang (*Ephedra sinica*) as a stimulant. Chemists have analyzed the constituents of these plants, and we now know that both of these traditional Chinese medicines (and many others) have some validity in terms of the active ingredients that are present.

The powdered root of Ch'ang Shan contains a number of chiral compounds displaying some antimalarial activity, including the compound drawn in Figure 5.1, which has the common name of β-dichroine. Although there has been some effort to determine whether this compound or one of the closely related compounds from this plant could be useful as a prescription medicine, they have been found to be too toxic. The most active ingredient in Ma Huang is the chiral compound

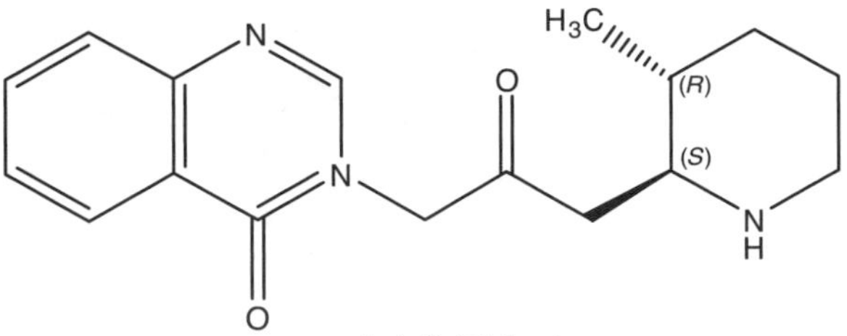

(+)-β-Dichroine

FIGURE 5.1. Molecular structure of β-dichroine, an active ingredient of the traditional Chinese medicine plant Ch'ang Shan.

ephedrine shown in Figure 5.2. This compound was used for medicinal purposes beginning in the 1920s, and until very recently was being used as a dietary supplement for weight loss reduction and athletic enhancement [4]. It was banned for use in the United States as a weight loss medication in 2003 by the FDA because it was considered to be too dangerous to the cardiovascular system, especially because of complications due to increased blood pressure and irregular heart rhythm.

The number of chiral compounds that have been isolated and identified from natural and folk medicines is very large, and major drug companies are still collecting and analyzing plants and animals from far corners of the world and then testing them in order to find active beneficial medicinal compounds. Living systems, of course, contain many different

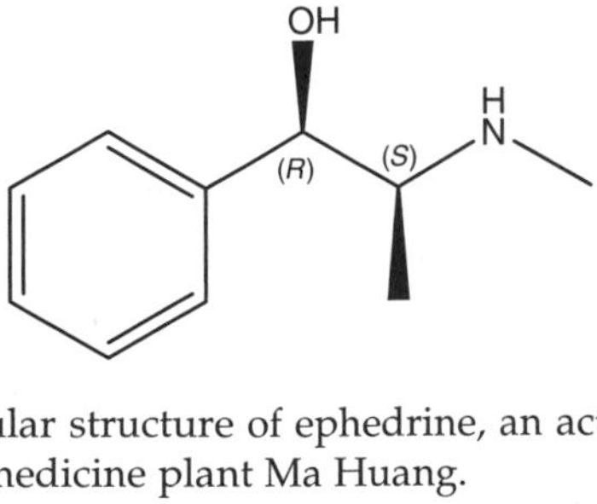

FIGURE 5.2. Molecular structure of ephedrine, an active ingredient in the traditional Chinese medicine plant Ma Huang.

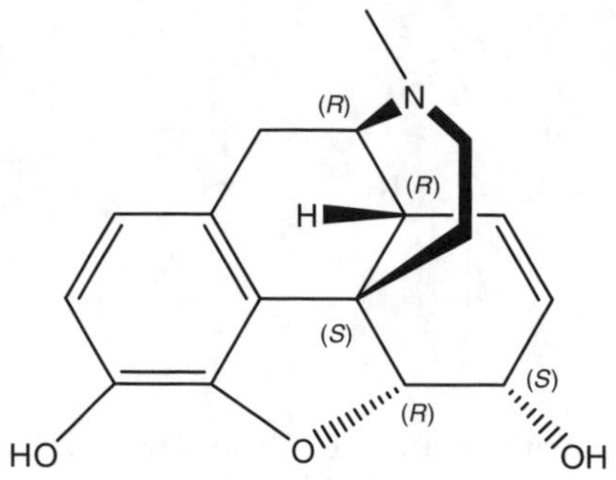

FIGURE 5.3. Molecular structure of morphine.

compounds, and it is often difficult to sort out from a complicated mixture what the active ingredients are, if there are any present at all. It is not surprising that one of the most widely known natural products that has been around for thousands of years is morphine. Morphine comes from opium powder, which is the dried juice from the unripe seed capsules of the poppy, *Papaver somniferum*. Morphine was purified from opium in the early 1800s, and as you can see from the structure in Figure 5.3, it contains five chiral carbon centers.

ENANTIOMERIC DIFFERENCES IN NATURAL PRODUCTS

Following the work of Louis Pasteur, Emil Fischer, and others in the second half of the nineteenth century, it became well known that many substances could exist as mirror-image isomers, and that compounds derived from living matter were almost always obtained as only one of the two possible forms. This knowledge led a number of scientists to investigate the differences between the physiological effects of mirror-image compounds. The Scottish scientist Arthur Cushny (Biographic Photo 5.1) began his study of the effect of different enantiomers

BIOGRAPHIC PHOTO 5.1. Arthur Cushny

on physiological activity after joining the faculty at the University of Michigan in 1893, and he continued this study when he moved to the University College London in 1903, and when he moved back to his native Scotland in 1918 to assume a professorship at the University of Edinburgh [5]. The late nineteenth century was a period in which much research was being performed on analysis of natural products, and Cushny began a series of experiments involving extracts from the "deadly nightshade" (*Atropa belladonna*) and related plants. Four of the compounds isolated from this plant are shown in Figure 5.4. Cushny was provided with some (−)-hyoscyamine (*l*-hyoscyamine) and a racemic mixture of this material, which is still called *atropine* today. Converting a substance to

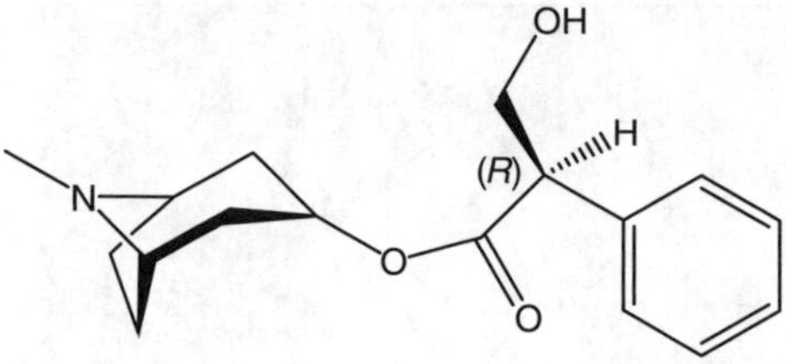

(−)-Hyoscyamine

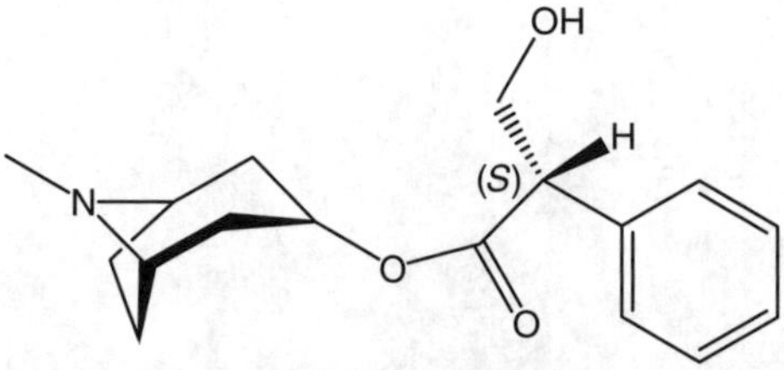

(+)-Hyoscyamine

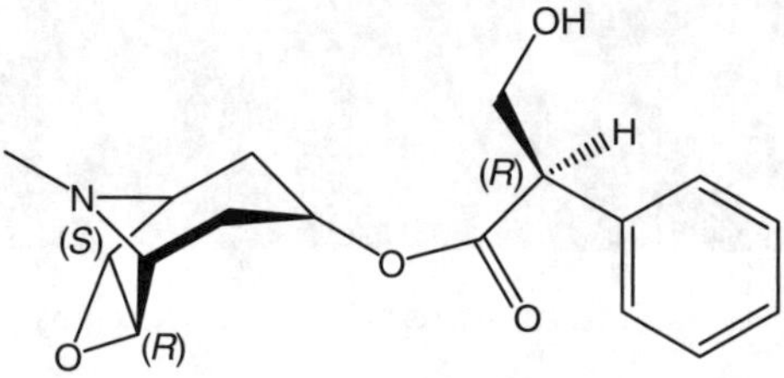

(+)-Scopolamine

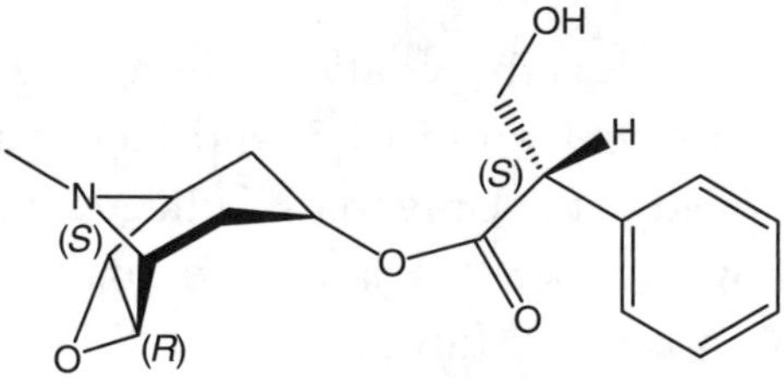

(−)-Scopolamine

FIGURE 5.4. Four active chiral compounds isolated from the "deadly night-shade" plant.

a racemic mixture is easier than preparing the enantiomer of the natural substance, or separating it from the racemic mixture. Cushny studied the effect of these two substances on frogs and some small mammals, and noticed that in some cases the effect of the pure (−)-hyoscyamine was twice that of the racemic mixture. From this result, he concluded that the active ingredient in the mixture was the (−)-enantiomer. He eventually was provided with a sample of (+)-hyoscyamine and determined that the effect of this compound was actually 20 times less than that of its enantiomer. Cushny is credited with the first clinical application of these enantiomeric differences when he published the use of (−)-hyoscine, commonly called *scopolamine*, to induce "twilight sleep" in patients [6]. The (+)-enantiomer was ineffective in this application. Among other uses, beginning in the early twentieth century and lasting for more than 50 years, scopolamine was administered to pregnant mothers just prior to child birth to put them into a semiconscious state so that they wouldn't remember the birthing process. This compound is still used today in alleviating motion sickness.

There were a number of related studies performed in the beginning of the twentieth century in which the pure naturally occurring compound was compared in effectiveness to the racemic mixture. For example, Fromherz demonstrated that natural (−)-adrenaline, which is secreted by the adrenal gland (see Figure 5.5), was 51–52% more active on blood pressure than is the racemic mixture, and thus concluded that (−)-adrenaline was 30–40 times more active than (+)-adrenaline [7]. This period of time saw tremendous advancements in chemistry, particularly in organic chemistry. Some of this effort and progress was no doubt driven by the need for modern weapons as the world was in and out of two devastating wars. By the midtwentieth century not only did chemists have the ability to extract, purify, and identify natural products much more efficiently and accurately, but an industry built around the synthesis of new compounds that did not occur in nature was prospering. Amid the many good consequences of the availability of new compounds were a few tragedies. One of

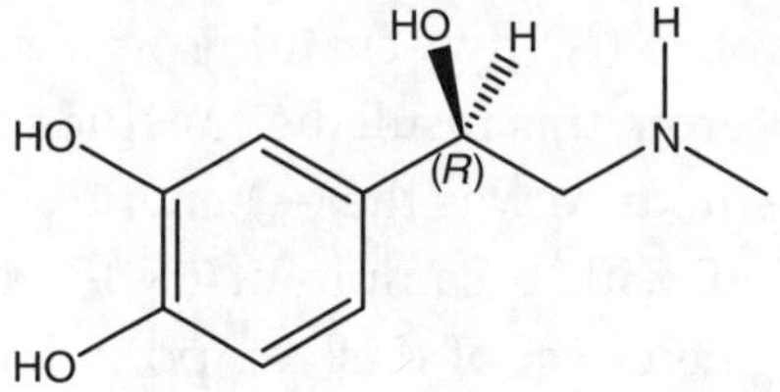

(–)-Adrenaline (or epinephrine)

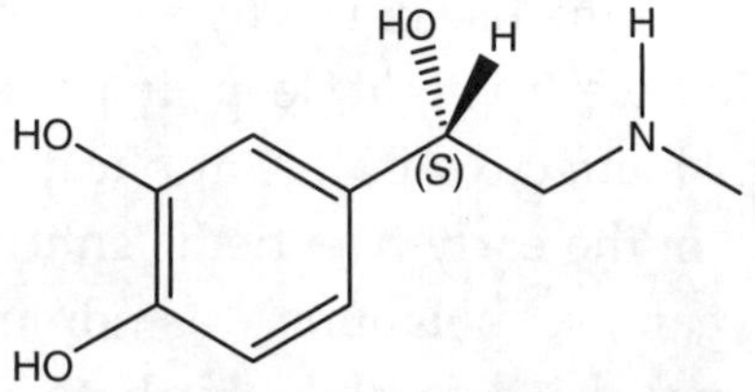

(+)-Adrenaline (or epinephrine)

FIGURE 5.5. Molecular structures of adrenaline enantiomers.

the worst was the development and marketing of the drug
thalidomide.

THALIDOMIDE AND REGULATION
OF CHIRAL DRUGS

Several books have been written on the story of thalidomide,
and the reader is referred to the suggested reading list at the end
of this chapter for much more information than will be given
here. Thalidomide is a chiral compound (see Figure 5.6),
and this fact, as we will see, played an important role in
this story.

The German chemical company Chemie Grünenthal produced a compound that they named *thalidomide* in the 1950s
while trying to find an inexpensive way to prepare an antibiotic
from proteins. This compound had no antibiotic activity, but
it didn't kill rodents, dogs, or cats when fed to them in large

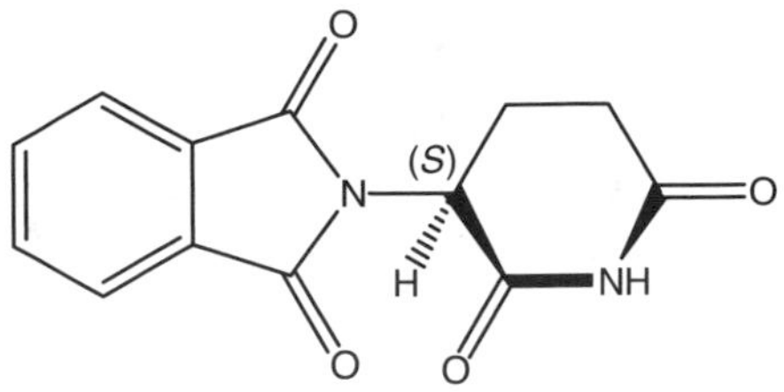

(S)-Thalidomide

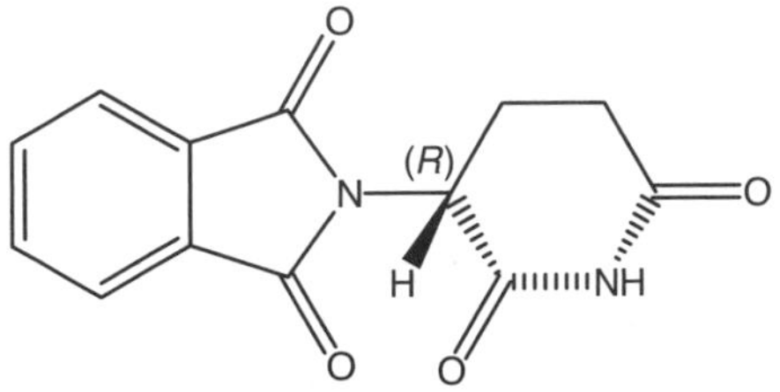

(R)-Thalidomide

FIGURE 5.6. Molecular structures of thalidomide enantiomers.

quantities, and since it didn't seem to have any other side effects, they looked around for some possible medical use. They apparently distributed the compound for free to physicians in Switzerland and Germany as a possible aid to people with seizures, and heard back that some patients experienced a deep sleep and a calming feeling. Amazingly, the drug soon became widely used to relieve the symptoms of morning sickness in pregnant women. This was all done without any significant testing. The tragedy is that this drug caused more than 10,000 birth defects worldwide. Fortunately, it never made it to the market in the United States because of a heroine, Frances Oldham Kelsey, in the FDA who refused to approve the drug without evidence that it was effective and safe [8]. Even so, it was distributed on an experimental basis in the United States in the early 1960s to approximately 20,000 patients, including several hundred pregnant women, and 17 thalidomide-caused birth defects were officially reported.

It has now been well established that the (R)-(+)-thalidomide enantiomer is responsible for the sedative effects of this drug, and that the (S)-(−)-enantiomer is responsible for causing the birth defects [9]. It might initially seem possible to still use this compound as a sedative if the enantiomers are separated, and to prescribe the active R-enantiomer alone (i.e., ensuring that the patient does not ingest the S-enantiomer). Well, it turns out that in the body there is a fairly rapid racemization (interconversion of the two enantiomers) of this compound, so that the harmful enantiomer cannot be eliminated [10].

Certainly, this story highlights the importance of rigorous testing of new drugs for effectiveness and safety, and also the importance of chirality in drug development. The drug thalidomide became associated with all that was wrong with the post–World War II pharmaceutical industry. Remarkably, the story of thalidomide, however, does not end here. As just about a last resort, the drug in the racemic form was given to a number of people with leprosy at a leprosy clinic in Israel, where it provided almost immediate pain relief, and also dramatically decreased the number and severity of skin lesions. Thalidomide was approved by the FDA for use in the treatment of leprosy, and for a number of years in the 1980s and 1990s was also used legally and illegally as a treatment for the acquired immune deficiency syndrome (AIDS). It has also been approved by the FDA for the treatment of multiple myeloma, a cancer associated with the immune system, and is under study for use in several other diseases [11]. Revenues for the sale of thalidomide for these new purposes totaled more than $300 million dollars in 2004.

Perhaps the most important consequence of the thalidomide tragedy was the legislation of much more stringent regulations for new-drug approval in the United States. In 1963 President Kennedy signed the Kefauver Harris amendment, which required that a new drug must be shown to be safe and effective before it would be approved for use. It further required informed consent of individuals before they were given experimental drugs. It took until 1992 for the FDA to require that in order to market a chiral drug as a racemic mixture, the drug company requesting approval must determine the pharmacological and toxicological activities of both enantiomers, and monitor the interconversion of enantiomers in animals and humans. Only in cases "where little difference is observed in activity and disposition of the enantiomers" will racemic mixtures be approved.

CHIRALITY AND DRUG ACTIVITY

The life cycle of a drug through the body involves absorption of the substance, transport and distribution, binding to receptors, metabolism, excretion, and other functions. All of these processes have the potential to be selective for different enantiomers; therefore, the number of chiral drugs in which the individual enantiomers have equal activity is relatively low [12]. More often the enantiomers of drugs have different quantitative effectiveness, and sometimes very different therapeutic properties, as was the case for thalidomide. Other situations exist, including cases in which the racemic mixture is more effective than either enantiomer alone because of some combined effects [13]. In this section, to illustrate the variety of situations that currently exist, we present the structure of a few common chiral drugs, emphasizing the similarities and differences in enantiomeric properties.

IBUPROFEN

The structure of ibuprofen is given in Figure 5.7, where we show both the R- and S-enantiomers. It is a nonsteroidal antiinflammatory drug (NSAID), and is marketed under various names,

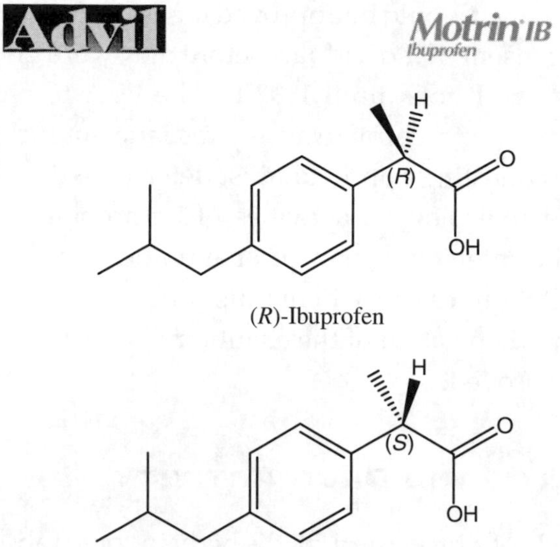

(*R*)-Ibuprofen

(*S*)-Ibuprofen

FIGURE 5.7. Molecular structure of ibuprofen enantiomers. (Advil® is a registered trademark of Wyeth Consumer Healthcare, and Motrin® is a registered trademark of McNeill-PPC Inc.)

including Advil® and Motrin®. It is known that *S*-ibuprofen is much more effective at relieving pain than is the *R*-ibuprofen enantiomer [14]. Although there has been much effort and success in preparing pure samples of the *S*-enantiomer, it was shown that, as we saw with thalidomide, ibuprofen racemizes in the body [15]. As a consequence, ibuprofen is almost always sold as a racemic mixture.

PROZAC

Prozac® is an oral psychotropic drug used principally to treat depression. In 2006 more than 21 million prescriptions were written by physicians for Prozac, making it the third most popular antidepressant medicine [16]. The common

PR⊙ZAC°
fluoxetine hydrochloride

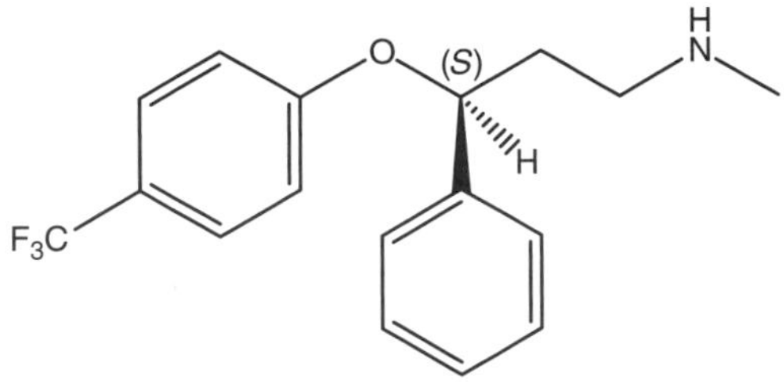

(R)-Fluoxetine

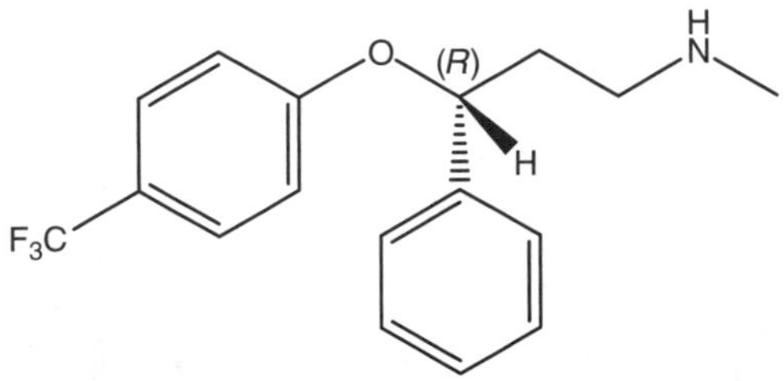

(S)-Fluoxetine

FIGURE 5.8. Molecular structure of fluoxetine enantiomers. (Prozac® is a registered trademark of Eli Lily and Company.)

chemical name for Prozac is fluoxetine, and as you can see from Figure 5.8, it has one chiral carbon center. Both enantiomers have been shown to be very active antidepressant agents; the only main difference is that the S-fluoxetine enantiomer is eliminated more slowly from the body, and is therefore the predominant enantiomer present in plasma. Prozac is also currently manufactured and sold as a racemic mixture.

CIALIS

Unlike its two main competitors, Viagra® (sildenafil) and Levitra® (vardenafil), the erectile dysfunction medicine Cialis®

is a chiral compound. The common (generic) chemical name for Cialis is tadalafil, and the substance that is prescribed is the (*R*,*R*)-enantiomer. In Figure 5.9 we show the structure of tadalafil and its two competitors. As you can see, tadalafil has

Cialis
tadalafil

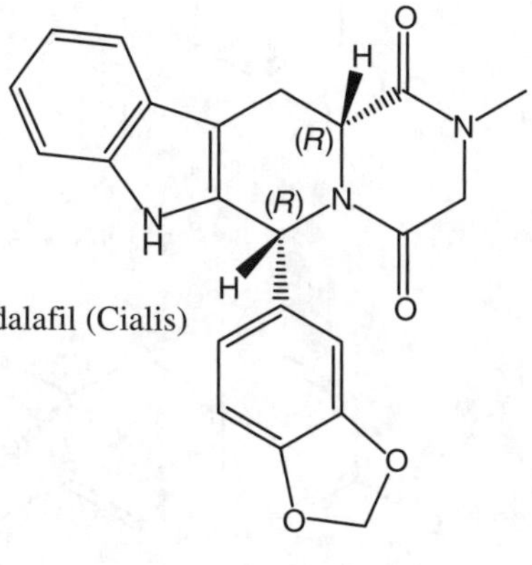

Tadalafil (Cialis)

VIAGRA
(sildenafil citrate) tablets

Sildenafil (Viagra)

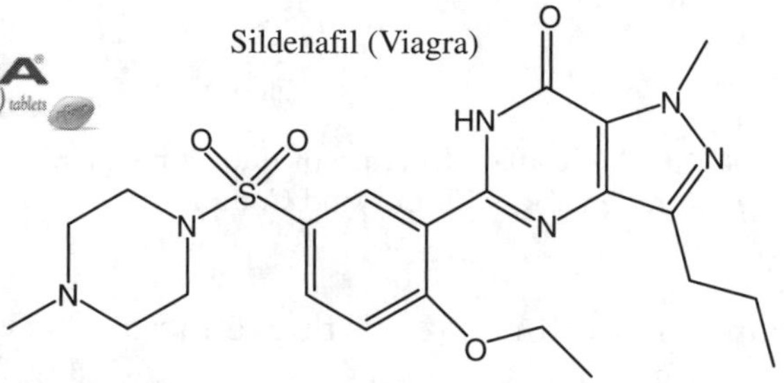

LEVITRA
(VARDENAFIL HCl)

Vardenafil (Levitra)

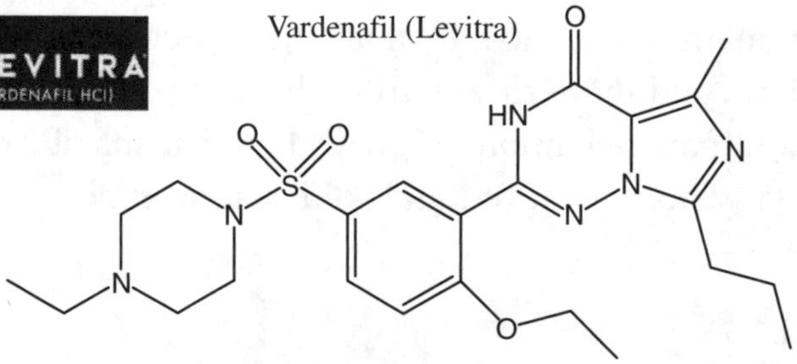

FIGURE 5.9. Molecular structures of tadalafil (Cialis® is a registered trademark of Eli Lily and Company), sildenafil (Viagra® is a registered trademark of Pfizer Inc.), and vardenafil (Levitra® is a registered trademark of Bayer Pharmaceuticals Corp.).

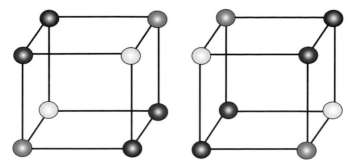

FIGURE 1.4. Mirror-image cubes with inversion symmetry.

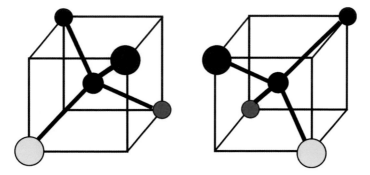

FIGURE 1.7. Tetrahedral enantiomers.

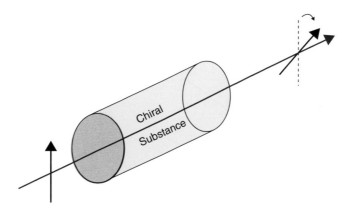

FIGURE 1.11. Schematic diagram of optical rotation.

SIDEBAR 1.E. Left-handed DNA.

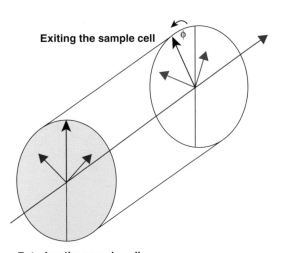

Exiting the sample cell

ϕ

Entering the sample cell

FIGURE 2.5. The rotation of plane polarized light. The counterclockwise rotation shown in this figure is denoted as $(-)$ and is called *levorotatory*.

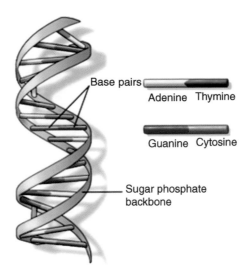

Base pairs

Adenine Thymine

Guanine Cytosine

Sugar phosphate
backbone

FIGURE 2.17. A three-dimensional model of a section of DNA. (Public-domain diagram from the US National Library of Medicine in the National Institutes of Health.)

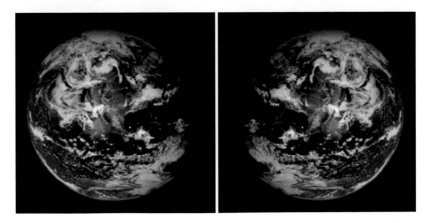

FIGURE 6.4. Earth from space, and its mirror image. (NASA, modified by the author.)

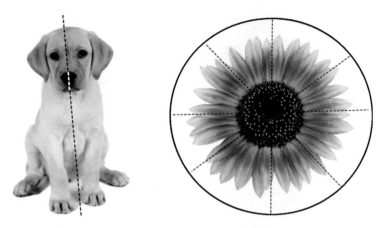

FIGURE 6.5. Planes of symmetry in plants and animals.

FIGURE 6.6. Two photographs of the Hawaiian hibiscus. (Photographed by Deb Shubat.)

Datura stamonium

Nerium oleander

FIGURE 6.10. *Nerium oleander* and *Datura stamonium*.

FIGURE 7.2. A prehistoric cave painting of a right hand.

FIGURE 8.4. The right-handed helical stairway at the University of Minnesota at Duluth. (Photographed by Brett Groehler.)

FIGURE 8.5. Attacking and defending the Swenson Science building at UMD! (Photographed by Brett Groehler.)

FIGURE 8.12. "De Valk" Windmill in Leiden, The Netherlands.

FIGURE 8.13. Windmills in Enkhuizen.

FIGURE 8.14. The blade of a Dutch Windmill in Arnhem with the canvas rolled up and secured.

two chiral centers. The discoverers of this drug prepared all four of the possible isomers *RR*, *RS*, *SS*, and *SR*, and tested each one separately for potency [17]. The (*R,R*)-isomer was almost 20 times more effective than the (*R,S*) compound and more than 1000 times more potent than the other two isomers. Cialis is sold as the (*R,R*)-enantiomer.

Nexium

Nexium® (esomeprazole) was introduced by the pharmaceutical company AstraZenaca as a replacement for the very popular Prilosec® (omeprazole) to control heartburn and acid reflux. Omeprazole is a racemic mixture and esomeprazole is the *S*-enantiomer (see Sidebar 5.A). The structures of the two enantiomers are shown in Figure 5.10. Note that in this molecule the chiral center is a sulfur atom, not a carbon. The structure around sulfur double bonded to oxygen (a sulfoxide) in this molecule is tetrahedral like carbon, but one of the tetrahedral coordination sites is composed of two electrons (not shown in this figure) that are not involved in bonding.

Sidebar 5.A

The Naming of Chiral Drugs. The nomenclature of chiral compounds is difficult and sometimes confusing. We have already introduced the prefixes *R* and *S*, (+) and (−), and D and L. We have already seen many common names for drugs that have an implied chirality, as in methamphetamine and ephedrine. The hyphenated prefixes *R/S*, D/L, and + /− render the alphabetic indexing of pharmaceutics problematic. One approach to this problem that has been used to help with this issue is to select drug names where the chirality has been imbedded into the common chemical name. One example that we have discussed is esomeprazole,

and there are also compounds that now begin with the letters
"ar" to indicate the *R*-enantiomer. This is useful, of course,
only when there is one chiral center. There are also some
existing names such as the amino acid <u>ar</u>genine, where the
prefix has nothing to do with the chirality of the carbon. More
common is the use of prefixes dex/dextro- or lev/levo- to
indicate the sign of the optical rotation for the specific
enantiomer being named. The prefix rac- is also sometimes
used to indicate that the substance being named is a racemic
mixture.

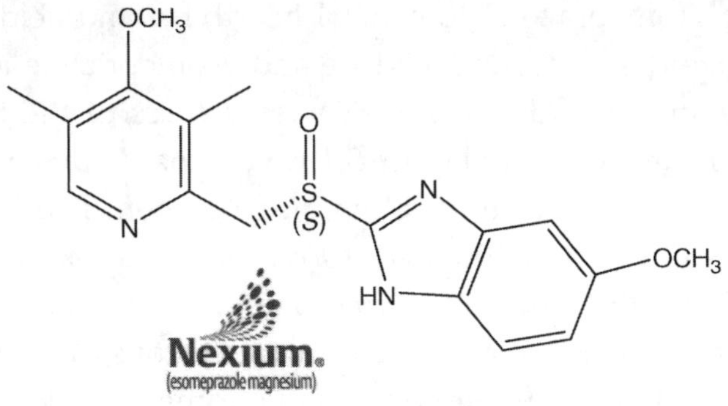

(*S*)-Omeprazole (esomeprazole)

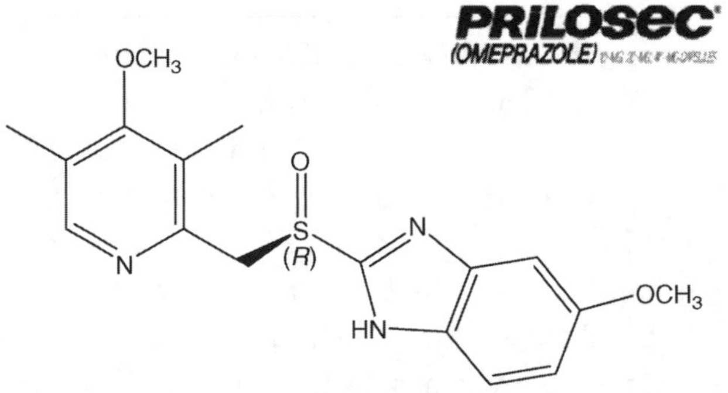

(*R*)-Omeprazole

FIGURE 5.10. Molecular structures of omeprazole enantiomers. (Prilosec®
and Nexium® are registered trademarks of the AstraZenaca group of
companies.)

The introduction and marketing of Nexium by AstraZenaca as a replacement for Prilosec has been the subject of some controversy. The timing of the patent application of the S-enantiomer of omeprazole corresponded closely to the expiration of the patent for the racemic mixture. Granting of the new patent would allow the company to continue to control the market for many years, if they could convince the physicians who write prescriptions for this drug that Nexium is a large improvement over Prilosec. It has been demonstrated that esomeprazole is much more effective than R-omeprazole in humans, and about twice as effective as the racemic mixture [18]. Interestingly, in rats it is the R-enantiomer that is more effective, and in dogs no difference between enantiomers was seen [19]. Since the racemic mixture contains only half of the amount by mass of the active ingredient, the result that the pure enantiomer is twice as effective is what one would expect. A decision for the various world patent agencies is whether there is enough new information to warrant a new patent. Even in cases where really new discoveries leading to the manufacture and use of one enantiomer are patentable, the advantages in safety or effectiveness of the enantiomer over the racemic mixture need to be evident. When a patent expires, other companies are allowed to make the compound and usually market it under the chemical name. Since these companies do not have near the expenses needed for research and development, and new-drug approval, the costs of the so-called generic drug are normally much lower than those of the original brand-name compound. In the case of Nexium or Prilosec, the generic racemic version, which may be much cheaper, may be just as effective at twice the dose. Because of the profit motivation and associated commercial aspects of the development of this drug, Malcolm Gladwell, writing in the *New Yorker*, said that "Nexium® has become a symbol of everything that is wrong with the pharmaceutical industry" [20].

LIPITOR

The cholesterol-reducing drug Lipitor® (chemical name *atorvastatin*) is the largest selling drug in the world. In 2006, sales of Lipitor in the United States were more than $12 billion dollars. The structure of atorvastatin is depicted in Figure 5.11. This drug was originally patented and sold as a racemic mixture of the (R,R)- and (S,S)-enantiomers. The patent on Lipitor' which is currently held by Pfizer, expired in Canada in 2007, and will expire in European counties in 2011. The Pfizer Company has tried to extend the patent for various reasons, and, of course, it would be in their financial interest to keep control of this drug away from generic manufacturers for as long as possible. As in the case of Nexium, Pfizer has applied for a new patent for the active enantiomer, which is the one shown in Figure 5.11, based in some part on a claim that the pure enantiomer was 10 times more effective than the racemic

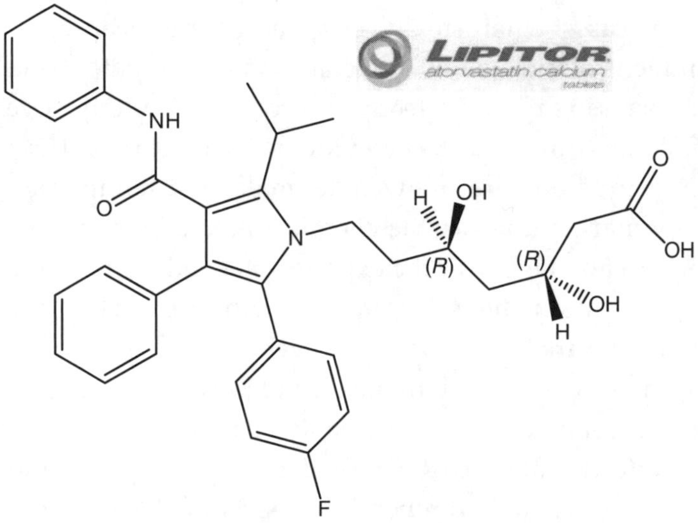

(R,R)-Atorvastatin

FIGURE 5.11. Molecular structure of the active enantiomer of atorvastatin. Lipitor® is a registered trademark of Pfizer Inc.)

mixture. As in the case of Nexium, one would normally expect the active enantiomer to be at most twice as effective as the racemic mixture. No explanation for this unexpected result was given in the patent application [21]. Previous publications had pointed to the importance of the (R,R) stereochemistry[22], so whether this was sufficient "nonobvious" information is open to question. Currently, this new patent has been granted in the United States, Spain, and Ecuador, but rejected in Canada, The Netherlands, and the United Kingdom.

CHIRAL SWITCHING

Changing the status of a commercial drug from a racemic mixture to a pure enantiomer has been termed *chiral switching*. As described above, there are many good reasons for making this switch. For patients, of course, it makes sense to take only the specific compound that has been shown to be effective in treating the disease. Chiral switching, of course, would not have helped resolve the problems caused by the (S)-enantiomer of thalidomide, due to the racemization of this compound in the body. There are other examples, however, of the noneffective enantiomer leading to cases that have benefited from the use of only one enantiomer. One case is the drug perhexiline (Figure 5.12), which is used to treat abnormal heart rhythms. [Note that for this compound, we do not yet know which absolute structure, R or S, is associated with the direction of rotation (+) or (−) of plane-polarized light.] This medication, when given as a racemic mixture, was fatal to several individuals, because (+)-perhexiline was metabolized much more slowly than (−)-perhexiline, and built up very high concentration levels [23]. Even if we can be convinced that the mirror-image enantiomer is not harmful, most of us would prefer not to put unnecessary substances into our bodies.

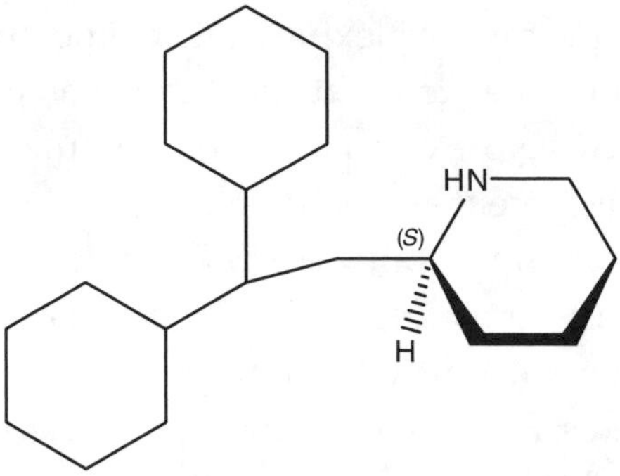

FIGURE 5.12. Molecular structure of (S)-perhexiline.

Chiral switching is also important to the pharmaceutical industry. If one is able to manufacture the active enantiomer efficiently, then there may be a savings in energy cost and waste. There is also the issue of patent protection and continuation. As was seen in some previous examples, if a company is able to prove that the enantiomer is more effective and safer than the racemic mixture, then a patent on the effective enantiomer, after marketing the racemic mixture, is a way to continue to reap large financial rewards. Success in obtaining a new patent on the enantiomer after a previous patent on the racemic mixture, however, is somewhat mixed [24]. In some cases a patent on an enantiomer, when the racemic mixture was already patented, was allowed, and in other situations it was deemed as "obvious from prior art" [19].

Sometimes it seems as though logic makes no sense when it comes to patent law, and there are some interesting cases involving chiral switching that highlight this aspect of the world in which we live. Perhaps the most interesting case involves the drug company Sepracor, which has been successful in getting patents on individual enantiomers of drugs that were already being used as racemic mixtures. US Patent 5,114,714 held by Sepracor claims that the R-enantiomers of

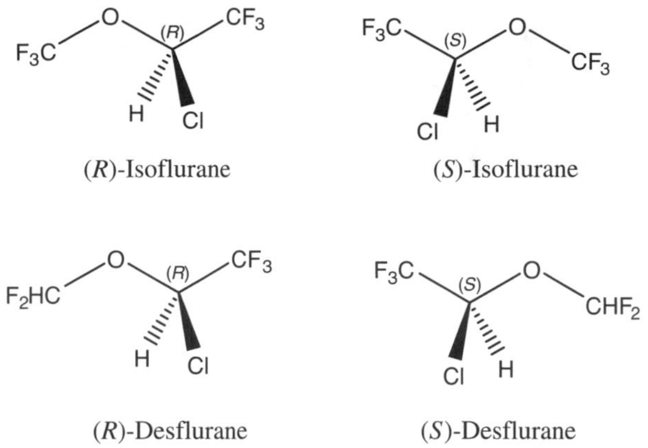

FIGURE 5.13. Molecular structures of the enantiomers of isoflurane and desflurane.

the anaesthetics isoflurane and desflurane (see Figure 5.13) are better than the racemic mixtures of these compounds, and US Patent 5,114,715, also held by Sepracor, claims that the S-enantiomers are better than the racemic mixtures. This company is clearly prepared for either possibility!

EPHEDRINE, PSEUDOEPHEDRINE, AND METHAMPHETAMINE

It should be obvious that the biological activity of drugs is directly related to the chiral structure of the compound being used for the particular medical purpose, and is often closely dependent on the chirality of the molecule at specific chiral centers. One of the best examples of this chiral selectivity is the different pharmacological properties of ephedrine and its diastereoisomers. As described above, the most active ingredient of the Ma Huang plant is (−)-ephedrine, which we show in Figure 5.2 and again in Figure 5.14 along with the other compounds with the same overall structure but different chirality at the two chiral carbon centers. The compound with the

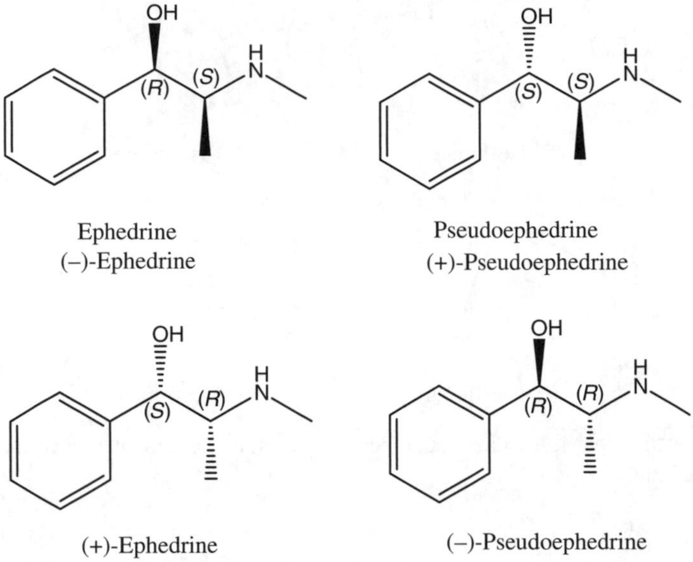

FIGURE 5.14. Molecular structures of ephedrine and related compounds.

common name pseudoephedrine is marketed by Pfizer under the name Sudafed, and it is the (1S,2S) isomer. Both (−)-ephedrine and (+)-pseudoephedrine are constituents of the Ma Huang plant, and are true *natural products*. The two compounds (+)-ephedrine and (−)-pseudoephedrine are predominantly obtained as synthetic enantiomers.

Whereas (−)-ephedrine was banned for use as a dietary supplement by the FDA, it was reported in a patent by Warner Lambert in 1999 that (+)-ephedrine may be useful as a decongestant and appetite suppressant without the side effects on the central nervous system (CNS) that is seen for the natural (−)-ephedrine enantiomer. Of course, Pseudoephedrine (Sudafed) is an over-the-counter medication in the United States widely used to treat nasal congestion associated with colds or the flu. The enantiomer (+)-pseudoephedrine has also been patented by Warner Lambert as a decongestant with fewer

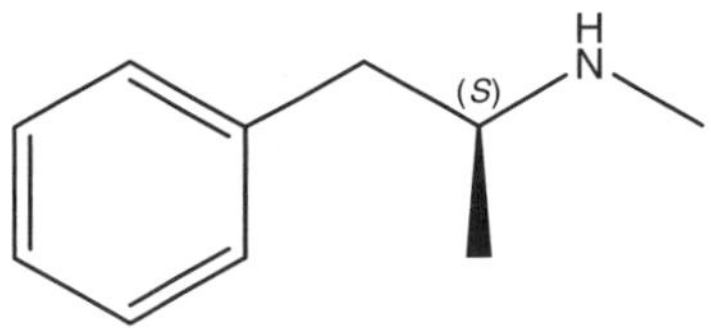

Methamphetamine

(+)-Methamphetamine

d-Methamphetamine

FIGURE 5.15. Molecular structure of methamphetamine.

CNS stimulatory effects, but it has not been made available for public use.

As every user of Sudafed in the United States knows, the purchase of this over-the-counter drug is now strictly regulated. The compound (+)-pseudoephedrine has a structure very similar to that of methamphetamine as shown in Figure 5.15. Methamphetamine is a potent CNS stimulant and a potent neurotoxin, and is a widely abused drug around the world. The similarity in chiral structure between these two compounds is the principal reason why the chemistry required to convert (+)-pseudoephedrine to (+)-methamphetamine is not difficult. This has led to many illegal "meth labs" and to far too many destroyed lives. The regulation of the purchase of Sudafed is to limit the availability of the starting material for the manufacture of methamphetamine. Because of the restrictions placed on pseudoephedrine, the most common over-the-counter decongestant is now the compound phenylephrine shown in Figure 5.16. This is sold, for example, under the name Sudafed PE, and is the decongestant contained in many other cold medicines. Note the similarity between the chiral structures of (−)-phenylephrine and (−)-ephedrine. Finally, (−)-methamphetamine, sometimes called *levomethamphetamine*, is also an over-the-counter decongestant, and is the active ingredient

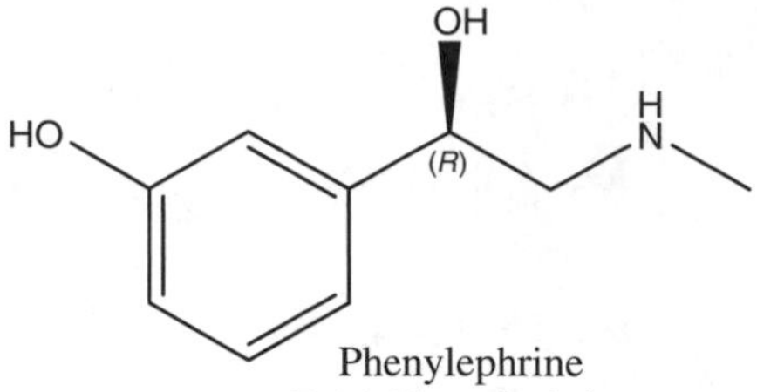

Phenylephrine
R-(−)-Phenylephrine

FIGURE 5.16. Molecular structure of phenylephrine.

in Vicks vapor inhaler in the United States. The use of this product by the Scottish Olympic skier Alain Baxter, however, led to his losing his silver medal in the 2002 Olympics. (see Sidebar 5.B).

Sidebar 5.B

2002 Olympic Drug Testing. The sports fans of England were overjoyed when Alain Baxter placed third in the slalom in the 2002 Olympic Games in Salt Lake City, and therefore walked away with the bronze medal. This was to be the first British medal in an Olympic alpine skiing event. Unfortunately, it was soon announced that Alain had failed a postrace drug test, and that specifically methamphetamine was detected in his urine [29]. He was shocked and argued vigorously that he had not taken any banned substance. It was soon discovered that Alain had used a Vicks Inhaler for a bad cold, not knowing that in the United States, this over-the counter product contains L-methamphetamine. The British form of this product does not contain this substance. Alain and the British Olympic Association appealed the decision to strip him of his medal because he had not taken the stimulant D-methamphetamine, but rather had inadvertently taken the enantiomer, which was a decongestant, not a strong stimulant, and therefore not performance-enhancing.

After 2 days of deliberation, the International Olympic Committee turned down his appeal and required him to return the medal. The existing policy on banned substances apparently did not recognize differences in chirality. The newest list of banned substances effective January 1, 2008 by the World Anti-Doping Agency (WADA) at least recognizes that D- and L-isomers exist, but in the section on stimulants it states that "All stimulants [including both their (D- & L-optical isomers where relevant] are prohibited" Drugs are usually detected in urine only by mass spectroscopy, which is insensitive to chirality, so it would not be possible to distinguish between enantiomers. Presumably, after a failed drug test, athletes could always say that they were taking the enantiomer, and it would be very difficult to prove whether they were actually cheating.

PREPARATION OF PURE ENANTIOMERS

Living systems figured out billions of years ago how to make pure enantiomers. Twenty-first-century chemists strive to achieve the efficiency and chiral selectivity shown by biological molecules, and although great progress has been made, we are not yet very close to this level of synthetic ability, usually referred to as *asymmetric synthesis*. One of the most important principles in the design of chemical processes aimed at producing chiral material is that you need to have something chiral to get something chiral. Now, it is true that in that very first pool or very first ocean, primitive forms of chiral life might have started completely randomly from achiral or racemic compounds, or possibly from a slight enantiomeric excess of amino acids or other chiral starting material. But in that situation time was not a constraint. Pharmaceutical chemists, medicinal chemists, and others need to make chiral compounds on a much shorter timescale, and to do so they need to work from the available chiral "pool" of starting material. Of course, the only reason why living systems are able to be so efficient in asymmetric synthesis is because they are themselves chiral.

The synthetic approach used by chemists who need to produce only a few milligrams of a chiral material for testing or analysis will often be very different from that of a company needing to manufacture kilograms or more for commercial distribution. In general, the more chiral centers there are in a molecule, the more complicated the process will be. Some of the drugs we introduced earlier in this chapter had one or two chiral centers, and the processes developed for their production can be relatively straightforward. Sometimes very important drugs with many more chiral centers are isolated from plants or animals, but are available in only small amounts. Their widespread use relies on chemists devising ways to make them artificially. The preparation of a large usually natural product molecule with all the correct chiral carbon atoms is called a *total synthesis*. An example is the cancer drug Taxol® (chemical name paclitaxel; see Figure 5.17), which was first isolated from the

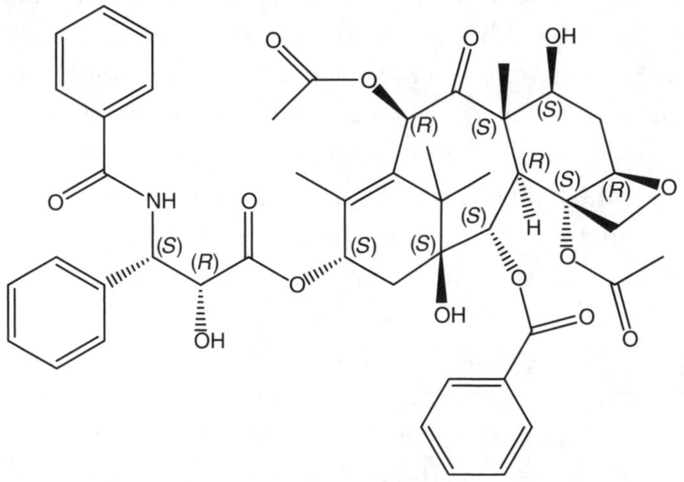

FIGURE 5.17. Molecular structure of paclitaxel.

bark of the Pacific yew tree [25]. It has 11 chiral carbon centers! The synthesis of this molecule was a major challenge. The first total synthesis, by Professor Robert A. Holton at Florida State University, started with the compound (−)-borneol. The synthesis was enhanced at a later stage by the availability of another larger natural product that served as a good starting point.

There are only a limited number of ways to prepare pure enantiomers. The most straightforward process is for the chemist to perform reactions on available chiral compounds that have been harvested from nature, namely "natural products." It is important in these chemical transformations that the integrity of the chiral centers or geometry be retained. Sometimes a chiral center may remain unchanged, with S staying as S and R staying as R, and sometimes in the specific reaction the chiral center may be inverted, with S changing to R and R changing to S. What must be avoided in this kind of synthesis is racemization of a chiral center. This was the approach of Emil Fischer (which we discussed in Chapter 2) when he was studying the various sugar molecules and relating them all back to glyceraldehyde. This is also the procedure used by Dr. Mori (presented in Chapter 4) in his efforts to determine the absolute structure of pheromones, and in the total synthesis of Taxol by Professor Holton and his collaborators.

Chemists have developed an extensive set of specific chemical reactions for which the enantiomeric properties are well known. In many cases, the absolute structure of the target natural product is not known, so chemists make a variety of compounds with different R or S chiral centers and compare various chemical and physical measurements such as optical rotation or nuclear magnetic resonance spectroscopy (NMR) from the synthesized compound and the natural product to see if they are the same. A second common approach is to synthesize racemic mixtures and develop a procedure to separate and purify the enantiomers. The separation process will necessarily have to involve some chiral material and make use of the differences in interaction between different enantiomers as we saw with the Pasteur separation of tartaric acid. Pasteur's approach was to dissolve the racemic tartrate negative ions derived from racemic tartaric acid with positively charged enantiomers of natural products. He observed that diastereomeric (i.e., R–R and R–S) salts formed as the solution cooled. One of the diastereomers crystallized before the other did because it was less soluble. These salts could be redissolved and the enantiomers of tartaric acid recovered. This type of diastereomeric salt separation is still used today in the laboratory and in larger-scale manufacture of enantiomers. One example is the separation of the S-enantiomer of the antidepressant drug Celexa® (chemical name citalopram) from the racemic mixture. The structure of this compound is given in Figure 5.18. The S-enantiomer is sold in the United States as Ciprolex® or Lexipro®. Although one needs a large amount of chiral material to form crystals with enantiomers of the substance that needs to be separated, a significant amount of this usually expensive chiral material can be recovered and used repeatedly [2].

A more recent technique for separating enantiomers has been the use of so-called chiral columns. In this approach a solution containing a racemic mixture is passed through a tube containing a solid chiral material. As the solution containing the enantiomers passes through the tube, either by gravity or through the action of a pump, the enantiomers interact differently with the chiral solid; this is called the *chiral stationary phase*. If, for example, the R-enantiomer is more attracted to the solid, or perhaps fits into chiral crevices in the solid a little better, then its velocity down the column will be slower than that of the S-enantiomer. At the end of the column, one has a detector (perhaps to measure optical rotation) to see when the separated

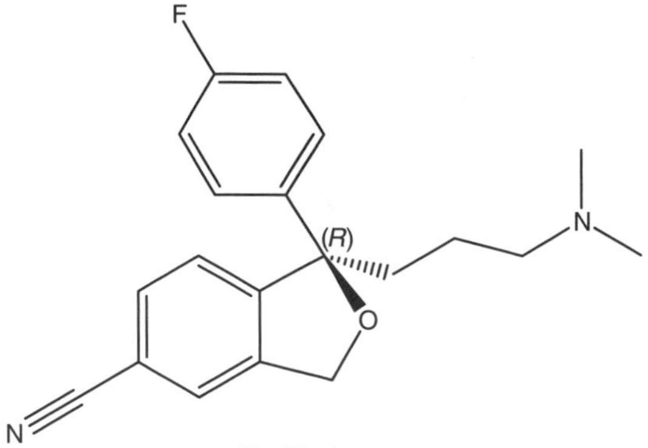

(R)-Citalopram

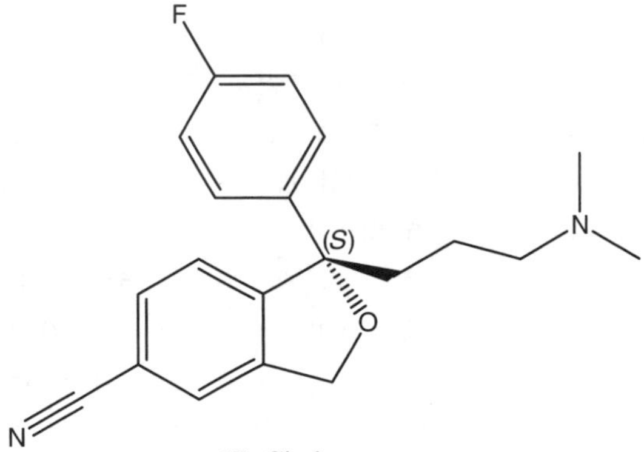

(S)-Citalopram

FIGURE 5.18. Molecular structure of S-citalopram.

enantiomers reach the bottom. In the best situation separate peaks are seen for each enantiomer. This is an example of chiral chromatography.

The use of chiral columns of various types has exploded since the 1980s, as the need for laboratory-scale and preparative-scale separation of enantiomers has increased. A number of chiral drugs are prepared for commercial purposes using chiral chromatography, for example, the antidepressant Zoloft® (chemical name S-sertraline) [26]. Its absolute structure is shown in Figure 5.19 [27].

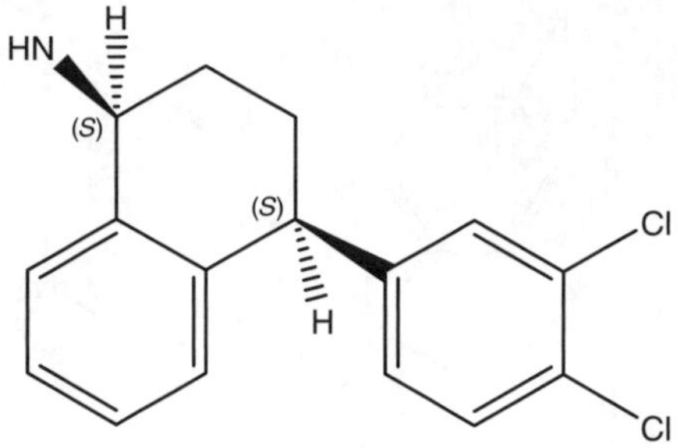

(S)-Sertraline

FIGURE 5.19. Molecular structure of S-sertraline.

Another very active area of asymmetric synthesis is the development and use of some type of reusable chiral catalyst. A *catalyst* is a compound that increases the rate (speed) of a chemical reaction. A chiral catalyst is able to increase the rate of formation of the desired chiral product. A good example is the chiral catalyst developed by Professor Barry Sharpless when he was a member of the faculty at the Massachusetts Institute of Technology. Along with William C. Knowles and Ryobi Noyori, he won a Nobel Prize in Chemistry in 2001 for the development of catalysts for asymmetric synthesis. The chiral catalyst that Sharpless developed is made up of the metal titanium (Ti) and a compound made from d-tartaric acid (the very compound that started it all!) The actual compound is the diethyl ester of d-tartaric acid [(−)-DET], which is shown in Figure 5.20. Using this catalyst, Sharpless was able to make the chiral compound R-glycidol from the achiral starting compounds shown in this figure. If the corresponding catalysis made from l-tartaric acid is used, the enantiomer S-glycidol is produced. Catalysts can be used repeatedly, so this is a way to prepare many chiral compounds from just a little bit of a chiral Ti(DET). This approach is becoming more widely used as chemists develop new types of chiral catalysts targeted at

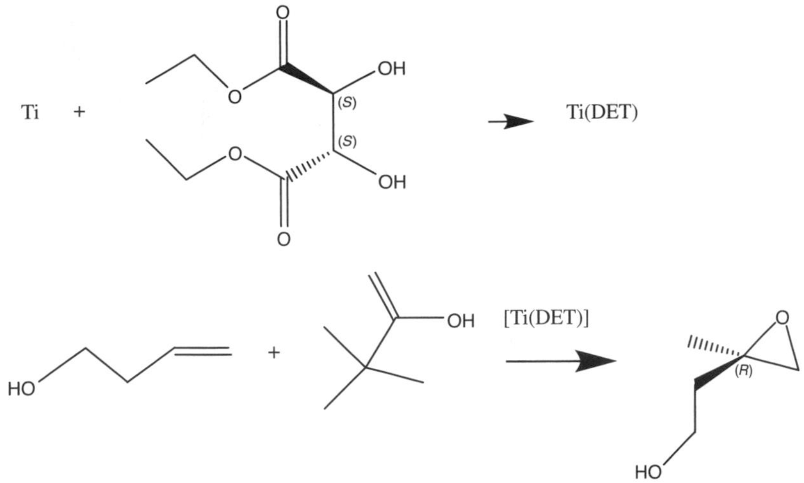

FIGURE 5.20. The synthesis and structure of Ti(DET) (DET = diethyl ester of d-tartaric acid) and example of its use in asymmetric synthesis.

conversion of specific achiral or racemic chemical structures to the desired enantiomer.

Modern molecular biology and biotechnology have also had a tremendous impact on the availability of chiral molecules used as starting points for asymmetric synthesis for pharmaceuticals, artificial sweeteners, cosmetics, and agricultural products. These industrial processes involve large-scale containers of living microorganisms that have been specially modified to produce the chiral products of choice from inexpensive food sources such as waste biomass from agricultural sources. For example, a number of commercially important L-amino acids have been manufactured from fermentation of sugar sources such as molasses by high-performance strains of *Escherichia coli* and other microorganisms for more than 50 years [28]. These industrial processes are very well developed. Specific mutant strains of microorganisms are used for different desired products, and precise protocols are used for the controlled

Ammonia   Fumaric acid   L-Aspartic acid

FIGURE 5.21. The synthesis of L-aspartic acid.

feeding and separation of the desired product from the fermenting culture. In addition to fermentation, a number of chiral amino acids are manufactured by the use of specific enzyme catalysts. An *enzyme* is a protein that serves as a catalyst for a specific chemical reaction in a living cell. Just like the catalysts described above, enzymes are not used up in a chemical reaction, but are used repeatedly. For example, L-aspartic acid is produced from achiral ammonia (NH_3) and fumaric acid (see Figure 5.21) by use of the enzyme aspartase from the bacteria *E. coli*. These industrial processes are big business. The artificial sweetener aspartame (Figure 5.22), which is made from L-aspartic acid (enzymatic production)

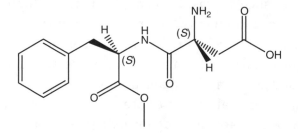

FIGURE 5.22. Molecular structure of aspartame.

and L-phenylalanine (fermentation) on a very large scale from biotechnology, has sales of more than $400 million dollars in the United States under the names NutraSweet® and Equal®.

SUMMARY

For practical, legal, safety, environmental, and other reasons, the importance of producing enantiomerically pure pharmaceuticals, food additives, and other chemicals as opposed to racemic mixtures is increasing. In the 50 years since the thalidomide tragedy, scientists have been able to develop a much better understanding of the differences in the biological effects of individual enantiomers. This has led to the development of new techniques for the production and separation of chiral materials. This trend will certainly continue, and in the near future a pharmaceutical or food additive or agricultural product will probably be used as a racemic mixture only rarely.

SUGGESTIONS FOR FURTHER READING

Burger, A., *Drugs and People; Medications, Their History, Origins, and the Way they Act*, Univ. Virginia Press, Charlottesville, 1988.

Brynner, R. and Stephens, T., *Dark Remedy: The Impact of Thalidomide and Its Revival as a Vital Medicine*, Basic Books, New York, 2001.

REFERENCES

1. Caner, H., E. Groner, L. Levy, and I. Agranat, Trends in the development of chiral drugs, *Drug Discov. Today* **9**(3):105–110 (2004).
2. Thayer, A. N., Centering on chirality, *Chem. Eng. News* **85**(32):11–19 (2007).
3. Burger, A., *Drugs and People*, Univ. Virginia Press, Charlottesville, 1988.
4. Newton, G. D., W. S. Pray, and N. G. Popovich, *J. Am. Pharm. Assoc.* **44**(2):211–225 (2004).

5. Parascandola, J., Arthur Cushny, optical isomerism, and the mechanism of drug action. *J. Hist. Biol.* **8**(2):145–165 (1975).

6. Cushny, A. R., *Biological Relation of Optically Isomeric Substances*, Balliere, Tindall and Cox, London, 1926.

7. Fromherz, K., The action of racemic and optically active adrenaline on the blood pressure, *Deutsche Med. Wochenschrift*, **49**:814 (1923).

8. Hilts, P. J., *Protecting America's Health: The FDS, Business, and One Hundred Years of Regulation*, Alfred A. Knopf, New York, 2003.

9. Blaschke, G., H. P. Kraft, K. Fickentscher, and F. Kohler, Chromatographic separation of racemic thalidomide and teratogenic activity of its enantiomers, *Arxneinmittel-Forschung* **29**(10):1640–1642 (1979).

10. Eriksson, T., S. Bjöurkman, B. Roth, A. Fyge, and P. Höuglund, Stereospecific determination, chiral inversion in vitro and pharmacokinetics in humans of the enantiomers of thalidomide, *Chirality* **7**(1):44–52 (2004).

11. Jacobson, J. M., Thalidomide: A remarkable comeback, *Expert Opin. Pharmacother.* **1**(4):849–863 (2000).

12. Triggle, D. J., Chirality in drug design and development, in *Chirality in Natural and Applied Science*, W. J. Lough and I. W. Wainer, eds., Blackwell Science, Ltd., Bodwin, Cornwall, UK, 2002, pp. 108–138.

13. Hutt, A. J., Drug Chirality and Its Pharmacological Consequences, in *Introduction to the Principles of Drug Design and Action*, J. Smith, ed., Harwood Academic Press, Amsterdam, 2005, pp. 117–183.

14. Hutt, A. J. and J. Caldwell, The importance of stereochemistry in the clinical pharmacokinetics of the 2-arylpropionic acid non steroidal anti-inflammatory drugs, *Clin. Pharmacokin.* **9**:317–373 (1984).

15. Caldwell, J., A. J. Hutt, and Y. Fasnel-Gigleux, The metabolic chiral inversion and dispositional enantioselectivity of the 2-aryl proprionic acids and their biological consequences, *Biochem. Pharmacol.* **37**:105–114 (1988).

16. Top selling generic drugs in 2006, Drug Topics, March 10, 2007.

17. Daugan, A., P. Grondin, C. Ruault, A.-C. Le Monnier de Gouville, H. Coste, J. M. Linget, J. Kirilovsky, F. Hyafil, and R. Labaudinière, The discovery of tadalafil: A novel and highly selective PDE5 inhibitor. 2:2,3,6,7,12,12a-hexahydropyrazino[1′,2′:1,6]pyrido[3,4-*b*]indole-1,4-dione analogues, *J. Med. Chem.* **46**: 4533–4542 (2003).

18. Hassan-Alin, M., T. Andersson, M. Niazi, and K. Roehss, A pharmacokinetic study comparing single and repeated oral doses of 20 mg and 40 mg omeprazole and its two optical isomers, *S*-omeprazole (esomeprazole)

and *R*-omeprazole, in healthy subjects, *Eur. J. Clin. Pharmacol.* **60**(11):779–884 (2005).

19. Rouhi, A. M., Chirality at work, *Chem. Eng. News* **81**(18):56–61 (2003).

20. Gladwell, M., High prices, *The New Yorker,* Oct 25, 2004.

21. Roth, B. D., US Patent, 5,273,995.

22. Roth, B. D. and W. H. Roark, Synthesis of a chiral synthon for the lactone portion of compactin and mevinolin, *Tetrahedron Lett.* **29** (11):1255–1258 (1988).

23. Gould, B. J., A. G. B. Amoah, and D. V. Parke, Stereoselective pharmaco-kinetics of perhexiline, *Xenobiotica* **16**(5):491–502 (1986).

24. Agranat, I. and H. Caner, Intellectual property and chirality of drugs, *Drug Discov. Today,* **4** (7):313–321 (1999).

25. Holton, R. A., H. B. Kim, C. Somoza, Liang, R. J. Biediger, P. D. Boatman, M. Shindo, C. C. Smith, and S. Kim, First total synthesis of taxol. 2. Completion of the *C* and *D* rings, *J. Am. Chem. Soc.* **116**(4):1599–1600 (1994).

26. Dapremont, O., F. Geiser, T. Zhang, S. S. Guhan, R. M. Guinn, and G. J. Quallich, *Process for the production of enantiomerically pure or optically enriched sertraline-tetralone using continuous chromatography.* US Patent 6,444,854 (2002).

27. Caruso, F., A. Besmer, and M. Rossi, The absolute configuration of sertraline (Zoloft) hydrochloride, *Acta Crystallograph.C: Crystal Struct. Commun.* **C55** (10):1712–1714 (1999).

28. Leuchtenberger, W., K. Huthmacher, and K. Drauz, Biotechnological production of amino acids and derivatives: Current status and prospects, *Appl. Microbiol. Biotechnol.* **69**:1–8 (2005).

29. British skier stripped of medal, *New York Times,* March 22, 2002.

LIST OF BIOGRAPHIC PHOTOGRAPHS, SIDEBARS, AND FIGURES

THE CHIRALITY OF LIVING SYSTEMS

By now the reader should be well aware of the fact that the most important constituents of all living systems are constructed from chiral molecules, and that these chiral molecules (amino acids, sugars, etc.) are almost exclusively of one chirality. We have also discussed the consequences of this homochirality on the helical structure of proteins and DNA, and showed that having the individual components of proteins, for example, being composed of only L-amino acids, leads to these larger structures also being predominantly of one helical chirality. At the molecular level it is relatively easy to see how the chiral monomers (L-amino acids) have combined to form chiral polymers (right-handed alpha helices in proteins). It is not so easy to see the impact of this chirality on the organization of structures and differentiation of functions that are necessary in the formation of living cells. Even more challenging is the effect of chirality, if there is any, on the shape or operation of much larger and much more complex organs and other components of a living system. In this chapter we will examine living systems in order to see if the chiral structures of our biomolecules have resulted in plants and animals that are chiral.

Mirror-Image Asymmetry: An Introduction to the Origin and Consequences of Chirality
by James P. Riehl
Copyright © 2010 John Wiley & Sons, Inc.

PRELIMINARY CAUTIONS

Before proceeding with this inspection of our world for chirality, we need to be mindful of several issues, namely, the integrity of photographs and other pictorial representations of chiral objects, the accuracy of classification of chiral objects, and the simple statistics associated with small sample sizes.

Before the age of computer-projected PowerPoint presentations, it was customary for scientists to prepare 35-mm slides for lectures and other presentations. Quite often a slide or two would be reversed (which was usually blamed on the graduate student assigned to run the slide projector) and the audience would see something like that pictured on the right side of Figure 6.1. If there is printing on the slide, the reversal is easily recognized and fixed, but if it is just a picture without any other clues, then you are looking at the mirror image of the true object without knowing that the picture has been reversed. Sometimes it is difficult to remember whether "Lady Liberty" is holding up the torch with her left or right hand! Back in the days when all

FIGURE 6.1. The Statue of Liberty and its mirror image.

photographs were printed from negatives, reversal of photographic prints also occurred. Although in a careful examination a photographer can identify the emulsion side of the negative, and therefore the side that should be "up," this was and is not always done, and when viewing photographs reproduced from historical or archival negatives, one must be on the lookout for mirror-image reversals. Obviously, if you are using such photographs to determine, for example, whether there are more left-handed baseball pitchers now than there were 50 years ago, you need to make sure that the negatives have not been flipped before printing. With modern digital cameras, it generally takes a mouse click to reverse an image, so accidental reversals should be much less common. However, unlike photographic negatives, pictures stored only on a computer as a "jpeg" (Joint Photographic Experts Group) or some other type of digital file cannot be picked up and examined to see if they have been reversed, and once the image has been saved as an inverted image, the integrity of the original image may be lost forever.

In some types of reproduction of images, mirror-image reflection is always a result. For example, whenever an image is transferred from a master image to a copy by some sort of printing or stamping process, the picture will be reversed. Thus lithographs, engravings, etchings, and other transfer processes are susceptible to misinterpretation of an absolute structure. It used to take real talent to produce a lithograph with text in the correct orientation, because it must have been made as a mirror image, and one needed to be able to write text as a mirror image. Nowadays transparent plastic sheets are used that enable the lithographer to draw pictures and add text in the normal orientation. Even in paintings from the sixteenth- and seventeenth-century one must be careful in interpreting the chirality of images, since some of the artists were aided with a device known as a *camera obscura*. The first examples of camera obscura were simply pinhole projection devices that

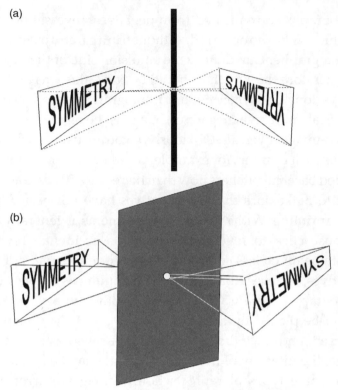

FIGURE 6.2. A "pinhole" camera obscura. In part (a) the image is projected onto a wall and traced; in part (b) the image is traced from a screen.

completely reversed the image onto a surface that the artist could trace. Depending on whether the image was traced or drawn from the back or the front, the image may be inverted as illustrated in Figure 6.2 (see also Sidebar 6.A).

Sidebar 6.A

Vermeer and the camera obscura. One of the most popular Dutch painters of the seventeenth century was Johannes Vermeer (1632–1675). Approximately 35 paintings are attributed to this artist, and he has developed an increasingly large number of devotees due to recent books and a popular movie based on a story inspired by one of his paintings, *Girl with a Pearl Earring* [16]. Some art historians downgrade his work, because they suspect that he used a camera obscura to trace the basic outlines of his subjects. Others do not believe at all that this diminishes the genius of Vermeer. An entire book has actually been written about his possible use of the camera obscura [17]. In addition to the perspective detail and other attributes of his paintings that lead one to the conclusion that he did use this device, the symmetry aspects of his paintings is also a subject of this study. In a simple pinhole camera obscura the image would be a mirror image, so that right-handed people would appear to be left-handed, whereas in more sophisticated devices involving lenses and/or mirrors the final images would not be reversed. It does appear that the women who are writing (see below) or drinking in Vermeer's paintings are, indeed, all portrayed as being right-handed, and therefore it appears that he was aided by the use of a camera obscura that did not invert the image.

"A Lady Writing" by Johannes Vermeer (National Gallery of Art, Washington, DC).

Obviously, if you are performing an analysis of photographs or other pictorial representations in order to determine whether there is an excess of one mirror-image form over the other, you need to be careful to ensure that the images that you are examining have not been reversed by the printing process. Clearly, it is much better to have access to the real objects than only to pictures of them. It is also important that categorization of the two mirror-image forms be understood and easily applied. From the discussions in Chapter 1 we noted that the connection between calling some object "right-handed" and the human right hand is not always obvious. Many objects have been categorized as being right- or left-handed on the basis of comparison with the right-hand screw or helix. These definitions are fairly rigorous, but as pointed out previously, there are numerous prominent examples where these definitions have been incorrectly applied.

Finally, we need to comment on summary conclusions based on an insufficient number of observations. We will focus in this chapter on examining our world and deciding whether certain objects or events are chiral or racemic. There is a direct analogy here to seeing whether, on flipping coins, we have more heads, more tails, or equal numbers of heads and tails. If you flip two coins, for example, the probability that you will get two heads is $\frac{1}{4}$ or 25% because there are only four outcomes, which are equally probable, namely, HH, HT, TH, and TT. If you flip N coins, what is the probability that you will get N heads? The formula that is used here is the probability $P = \left(\frac{1}{2}\right)^N$. So with 10 coins the probability that they will all land as heads (or tails) is $\left(\frac{1}{2}\right)^{10} = \frac{1}{1054} = 0.00098$ or 0.098%. If you are observing our world and you see 10 independent objects or events that could be either A or B, and they are all A, you would probably conclude that the probability of A and B occurring is not equal.

Let's do a real example. You walk to a baseball field and observe a team of nine players practicing. All nine are throwing the ball around with their right hand. If being a right- or left-handed pitcher (thrower) were equally probable, then the probability of seeing nine right-handed throwers would be $\left(\frac{1}{2}\right)^9 = \frac{1}{512} = 0.0019 = 0.19\%$. You would conclude from your observation of these baseball players that the probability of being a right- or left-handed pitcher is not the same. This is, of course, correct. Suppose that you observed two apes at the zoo and they both appeared to be using their right hand to peel a banana. Should you conclude that apes are predominantly right-handed? Clearly, the larger the sample size (N), the more confidence that you will have in your conclusions. It is also very important to be careful in determining any bias in the observations. This will be an important part of the discussion that follows this section. For example, returning to the game of American baseball, if for some reason you have access only to the data for players who play first base, you will see that about 30% of first-base players in the major leagues throw (pitch) with their left hand. Should you conclude that 30% of humans are left-handed? The answer is "no". Since left-hand pitchers in baseball do not play certain positions on the field, more than the average number end up at first base.

CHIRALITY IN NATURE

Should we be surprised to find chiral structures in nature? Surely we should not, because all living things are made up of chiral amino acids, sugars, and other components. One major questions is whether the fact that the building blocks of life are chiral controls the overall chirality of the macroscopic system, or whether this macroscopic chirality has a more macroscopic origin. We have already discussed the many possible chiral influences on the origin of the first chiral molecules, and in the context of this chapter we need to consider the same environmental influences on the possible chirality of living systems. As we know, every location on earth presents a living creature with a local environment to live in that is chiral. The combination of the spinning earth and gravity results in a weak helical Coriolis force everywhere on earth except at the equator. Although the net effect of the Coriolis force would be zero if averaged over the whole earth, in specific geographic locations where a particular species is evolving, this helical force is present. The Coriolis force is the origin of the urban myth concerning the direction of toilet flushing! (See Sidebar 6.B.)

Sidebar 6.B

Down the Drain. How often have you heard that that toilets in the northern hemisphere drain counterclockwise, but those in the southern hemisphere drain clockwise...or was it that in the northern hemisphere they drain clockwise and in the southern hemisphere they go counterclockwise? No one seems to remember which way it is, and, of course, no one seems to check this accepted scientific "fact." As in many other "scientific myths," there is some real science here, but very few people have the initiative to check out this commonly believed principle by looking at real drains. If you start filling up wash basins and watching how they drain, and keep track of whether the vortex formed in the drains is clockwise or counterclockwise, after examining a fair number of drains you will see that it is 50 : 50. The physics of this phenomenon is actually pretty easy to understand, if you can comprehend the *Coriolis force* or the *Coriolis effect*. The earth rotates from west to east, and the speed of a spot on the earth is faster, the closer you get to the equator. So if you think about shooting a cannonball directly south, it will appear to veer to the west, because the earth is turning faster where the cannonball lands than where the cannon itself was located. If you shoot the cannonball north, then the ball will appear to veer to the east because the launching spot is turning more rapidly than the landing spot. So all motions in the northern hemisphere appear to veer to the right to an observer on the earth, and similarly all motions appear to veer to the left in the southern hemisphere. This is the Coriolis effect, and is represented by the curved arrows in the figure drawn below.

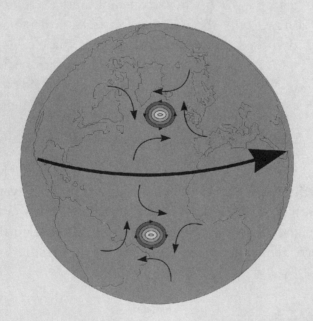

Now think of a large vat of water in which you have pulled the plug located in the bottom of the vat at the center. Water starts rushing from all directions into the center of the vat because water is falling out the bottom. Motions in the northern hemisphere turn to the right, and the net effect as shown in the figure is a counterclockwise vortex. The exact opposite behavior is observed in the southern hemisphere.

So theoretically, drains should behave differently in the northern and southern hemispheres. Why don't we observe this in reality? The reason is that the Coriolis force is very, very small. The force of gravity is 10 million times stronger [18]! Only under very carefully controlled conditions has this effect on drains actually been observed [19]. These scientific observations were made on carefully constructed vats of water with perfectly cylindrical drainpipes that were unplugged without disturbing the water. The scientists also found it necessary to eliminate any air currents in the room, keep the temperature constant, and wait for more than 24 hours after the vats were filled in order to see the expected vortex formed. So, we don't see these differences in drain vortices because our drains are far from perfect, and the way we fill the sinks imparts a clockwise or counterclockwise circulation that the liquid water "remembers" for a long time.

Another possible influence might result from the fact that plants, that are trying to follow the sun as they grow vertically, see a path in the sky that is left to right as they face south in the northern hemisphere and right to left as they face north in the southern hemisphere (see Figure 6.3). This results in a net

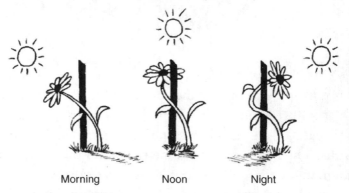

Morning   Noon   Night

FIGURE 6.3. A twining plant growing as a left-handed helix in the northern hemisphere. (Illustrated by Rachel MaKarrall.)

left-handed helical environment in the north and a right-handed helical environment south of the equator.

As we mentioned briefly in Chapter 3, the sunlight striking the earth is partially circularly polarized because of the influence of the earth's magnetic field and other extraterrestrial and terrestrial factors. Different locations on the earth see different levels and directions of circular polarization, but locally the environment would be chiral. Local geologic formations may be chiral, and this might influence the evolution of a species specific to that region. Furthermore, there are normal regional circulating weather patterns with prevailing wind directions and water currents that circulate in a clockwise or counterclockwise direction depending on the geometry of the coastline, prevailing wind directions, or other factors. In fact, the earth itself is chiral, as shown in Figure 6.4!

In a way, we should be surprised that there is so much symmetry in the natural world! The primitive geometry of life started out as spherical, but as soon as single-celled organisms started to form colonies to specialize functions, spherical symmetry was replaced by radial symmetry in the case of plants, and bilateral symmetry for animals. Plants need to distinguish

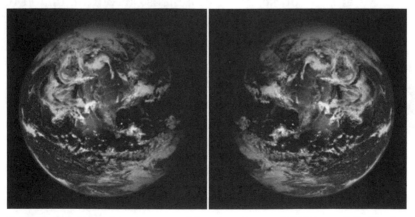

FIGURE 6.4. Earth from space, and its mirror image. (NASA, modified by the author.) See color insert.

FIGURE 6.5. Planes of symmetry in plants and animals. See color insert.

top from bottom, and leaves and branches from roots, and animals need to know where they are going, so their front has different organs and structures than their rear. All animals, except perhaps microscopic organisms, are influenced by gravity, and display differences in both top and bottom and front and back. In terms of symmetry elements that prohibit chirality, on the average, plants have an infinite number of vertical planes of symmetry, and animals in this simplified picture have one vertical plane as shown in Figure 6.5.

We are, of course, being quite loose in our definition of planes of symmetry. We are ignoring the fact that within the cells there are L-amino acids, D-sugars, and other compounds that, on mirror-image reflection, would convert to D-amino acids, L-sugars, and other enantiomers. This is in addition to the chirality of atomic nuclei described briefly in Chapter 3. We are also thinking of an "averaged" plant. Branches and roots, of course, are equally likely to appear at every position around a tree, so if we looked at all the trees of this type we would get an even distribution of branches in all directions. Trees and other plants don't have a back, front, left, or right, only a top and a bottom. Our view of animal and plant symmetry so far in this discussion has been limited to external structure, but we know that most animals, including humans, have an asymmetric distribution of internal organs, and the individual components of plants such as roots and leaves do not possess radial symmetry.

In this chapter we will not reproduce the several more recent informative summaries in the literature on the asymmetry of plants and animals, but in the following sections discuss a few examples. The interested reader is referred to a booklet written by A. C. Neville, who collected observations on asymmetry from bacteria, insects, mollusks, reptiles, birds, mammals, and other species [1]. Also, Martin Gardner has included two chapters on plant and animal asymmetry in his book *The New Ambidextrous Universe* [2] and more recently two other book chapters have been published on aspects of asymmetry in nature [3,4]. These sources also contain many references to primary-source articles about the observation of animal and plant asymmetry.

CLASSIFICATION OF PLANT AND ANIMAL CHIRALITY (ASYMMETRY)

In examining the external structure of living things for asymmetry, we shall ignore "normal" differences in coloring, slight species-dependent deformations or imperfections, or other differences that do not seem to be genetics-based asymmetric variations. In the case of plants we will apply the radial averaging described above. These considerations are consistent with previous treatments of this topic. Of course, quite subtle judgments on what appears to be genetics-based variations and normal variations are needed, and these judgments are certainly open for criticism. In the discussion presented below, we have concerned ourselves with discussing a few examples in which the asymmetries are obvious.

For purposes of this book we classify the external asymmetry found in nature into the following categories:

Symmetric—no evident asymmetry or chirality exists.

Racemic—the species displays asymmetry, but the entire population is composed of exactly 50% of each mirror-image form.

Asymmetric—there exists a statistically significant preference of one mirror-image form over the other.

These definitions are somewhat different from those of Palmer and others [4], and are based on the previous discussions and definitions in this book of chirality and asymmetry in molecules.

SYMMETRIC LIFE

Most of life that we see is symmetric. If tomorrow morning we wake up in a forest that was inverted into its mirror image while we were asleep, we probably wouldn't notice it. At first glance, the trees look the same, the ants, flies, and mosquitoes look the same; the trout, minnows, and frogs look all right; and the bears, deer, squirrels, mice, and birds all appear to us to be normal. It is certainly true that at this level of observation we will have to look hard in this forest to find something growing, crawling, walking, climbing, swimming, or flying that is not symmetric.

RACEMIC LIFE

After sitting for a while in this forest, you may be ready to conclude that life is all symmetric—that is, until you happen to see some flowers growing nearby. If this forest happens to be in Hawaii, the flowers might be the Hawaiian hibiscus, the state

FIGURE 6.6. Two photographs of the Hawaiian hibiscus. (Photographed by Deb Shubat.) See color insert.

flower of Hawaii (*Hibiscus brackenridgei*). As you contemplate your situation, you notice that the petals on these flowers are, indeed, chiral. The petals on the flower on the left in Figure 6.6 are arranged such that as you go clockwise around the flower, the individual petals overlap the next petal, whereas the petals on the flower on the right side of this figure overlap in the opposite direction. We describe the arrangement of petals on the left side as left-handed or sinistral, because of the comparison to a left-handed helix, and those on the flower on the right as dextral, due to the comparison with a right-handed helix. By our definition, the Hawaiian hibiscus flower is an example of a racemic species. It has been observed that plant species of the families Guttiferae, Malvaceae, and Oxalidaceae generally produce flowers with an equal number of the two mirror-image forms [5].

There are also a number of animal species that display obvious asymmetry in the individual, but when a large number of animals of the species are examined, the probability of the particular feature is random, and, therefore, occurs in exactly one-half (50%) of the population. A good example is the male

FIGURE 6.7. A fiddler crab.

fiddler crab pictured in Figure 6.7. As you can see, this species of crab and many other species, some in one sex and some in both sexes, have one large claw for the more aggressive behavior of food catching and defense, and a much smaller claw that they use for more delicate tasks such as eating [1]. Which claw is larger or smaller is completely random. Presumably crabs developed a larger claw because of the evolutionary advantage it provided in the functions of eating, mating, defense, and related activities. Since these larger claws are on either side, one can conclude, perhaps, that the world in which the crab evolved was not one with significant asymmetry. The references cited above describe additional examples of racemic species from bacteria, insects, birds, amphibians, and larger animals.

ASYMMETRIC LIFE

Back in this mirror-image forest, you have just about figured it out, or so you think! Sure, there exists some asymmetry in nature, like the Hawaiian hibiscus and the fiddler crab, but this mirror-image forest still looks the same as it did before this magical reversal, because we have found just as many sinistral as dextral forms. But wait, you suddenly come across a large number of honeysuckle plants, and a large number of snails. The honeysuckle plants are growing vertically all in the same

FIGURE 6.8. A left-handed helical honeysuckle stalk. (Photographed by Deb Shubat.)

helical direction, and the snail shells all seem to be of the same chirality!

Indeed, some animals and plants appear to be overwhelmingly asymmetric. Honeysuckle plants (genus *Lonicera*) grow vertically in a left-handed helix as shown in Figure 6.8, as does hop, whereas plants such as bindweed and runner bean grow as right-handed helices [5]. There are many other examples of this behavior, and one simply needs to explore one's garden to find other examples. What is the origin of this chirality? There has been some speculation that the direction of "twining" was influenced by the direction of the sun's movement. In Figure 6.3 we illustrated the fact that a plant growing vertically in the northern hemisphere, that wanted to face the sun as much as possible, might be expected to follow the path of a left-handed helix as it grew. The opposite helicity would be expected for a plant in the southern hemisphere. Another possibility might be for the direction of helicity to be influenced by the weak Coriolis force. The Coriolis force influences the direction of low- and high pressure areas in the northern hemisphere, which are exactly opposite to those in the southern hemisphere. There are also circulating patterns of air and water masses in the northern hemisphere that vary with latitude, and that are, on average, opposite to the air and water patterns in the southern hemisphere. Adding to this motion a vertical growth movement, we might expect helical influences in northern hemisphere latitudes that would be opposite to those seen in the corresponding southern hemisphere latitude.

Some biologists have considered gravity to be a requirement for this type of helical growth; however, in an interesting experiment involving sunflower plant seedlings on the Spacelab I mission of the space shuttle in 1984, this helical growth was shown to still be present in the microgravitational environment of low-orbit space flight [6]. Moreover, in a study of 1485 twining plants selected by straight line walking until about 100 twines were found in 17 sites spanning both hemispheres and nine countries, it was determined that 92% of the twines that were examined grew in right-handed helices, and there was no correlation with location with regard to hemisphere or Coriolis force [7].

The fact that honeysuckle and other plants consistently grow in one helical form, and that the direction of helical growth is not influenced by gravity, Coriolis force, or the path of the sun, really points to genetics as the basis for this asymmetric growth pattern. The development of new techniques in molecular and cellular biology has allowed for great progress in developing an understanding of plant growth, and there have been some new insights obtained on helical growth [8,9].

FIGURE 6.9. The plant *Arabidopsis thaliana*. (Photographed by Deb Shubat.)

For example, by irradiating seedlings of the plant *Arabidopsis thaliana* (Figure 6.9), mutants can be made that grow in either a left- or right-handed helix. This plant is often used as a model for this type of research, since it is small, has a rapid life cycle (the germination–seed cycle takes about 6 weeks in the laboratory), and it was the first plant to have its genome sequenced. Researchers have been able to identify specific genes involved in controlling the direction of helical growth of this plant [9]. They also determined their location on specific chromosomes that differ in the two mutants, and a model has been developed to understand the mechanism by which this plant selects the helical direction of growth. For our purposes, the most interesting result is, in fact, that the helical twining of certain plants is controlled by the plant's genes. Although programming into a plant's gene the ability to climb to seek sunlight is obviously an important trait, there doesn't appear to be any specific advantage to climbing in a left or right helical fashion. Perhaps, the effects of sun movement, weather patterns, and other factors influenced the helicity at a very early stage of evolution of this type of plant.

There are other examples of asymmetry in plants that seem to be genetically based. It has been reported that the petals of the flower of *Nerium oleander*, shown in Figure 6.10, for example, are arranged in a right-handed helical pattern, and those of *Datura stamonium*, also shown in this figure, form a left-handed helical

Datura stamonium

Nerium oleander

FIGURE 6.10. *Nerium oleander* and *Datura stamonium*. See color insert.

FIGURE 6.11. The snail *Helix pomatia*.

pattern. Sometimes these helical patterns are hard to see in photographs, and in order to really verify for yourself the asymmetric nature of certain flowers, you need to go to your garden or the local botanical garden at the right time of year.

How about those snails that we saw with all the same shell chirality? In Figure 6.11 we show a photograph of the most common snail (escargot) served in French restaurants, the *Helix pomatia*. Also in Figure 6.12 we show views of the *Helix pomatia* empty shell. The nomenclature of snail shells follows that used previously. As you proceed into the shell pictured in Figure 6.12, your path turns in a clockwise direction. This is a right-handed helix, and this shell is dextral. This, again, may be difficult to see in a photograph; however, if you look at the snail with the point of the shell away from you, you can easily

FIGURE 6.12. Snail shells for *Helix pomatia*.

recognize a dextral snail by the increasing radius of the clockwise swirl of the shell as pictured in Figure 6.12. To obtain information about snail shell chirality directly from the source, we visited a snail farm in the Burgundy region of France, and we asked the snail farmer how often he sees a sinistral snail. His response for another snail served in French restaurants, *Helix aspersa maxima*, was "Un sur un million" (one in a million). Another estimate that is no doubt based on more hard data put the probability of a *H. pomatia* being sinistral at 1 in 20,000 [3]. Although rare, sinistral snails do exist, as shown in Figure 6.13 for the species *H. aspersa maxima*.

Not only are *Helix pomatia* and *Helix aspersa maxima* predominantly dextral, but so are most of the more than 70,000 species of gastropods. Overall it is estimated that more than 90% of all gastropod snails are dextral [10]. As a result, sinistral shells of many species are quite prized possessions. There are, however, species of gastropods that are predominantly sinistral, and some with an approximately equal number of dextral and sinister individuals [11]. The question as to whether there was some evolutionary advantage to a gastropod to be dextral has been considered, but no sensible explanation has yet been put forward. There are so few examples of such asymmetry in the animal kingdom, that this question remains an interesting one.

FIGURE 6.13. Enantiomeric snail shells for *Helix aspersa maxima*. (Photographed by C. S. Riehl.)

So, why are gastropods so predominantly dextral? Since the 1920s it has been known that the chirality of snails is an inherited characteristic [12]. A detailed understanding of the genetics has been complicated by the fact that the expression of the gene that determines the chirality is delayed by a generation. The chirality of the offspring is dictated by the genotype of the mother, but the exact way that she passes on this information is not known. Also, for many snail species, including *Helix pomatia*, mating between individuals of opposite chirality is virtually impossible because of the physical location of the reproductive organs [13]. There is still much to be learned about the development of snail chirality, and although probably not the most important area of biology, these chiral characteristics do provide geneticists and developmental biologists a window into the evolutionary biology of delayed inheritance.

There are other examples of asymmetry among animals, but some of these do not seem to be ascribed simply to genetics. For example, flounder are born symmetric, but settle on the bottom of the ocean after a few weeks. In a few more weeks, one of the flounder's two eyes moves from the underside to the topside (see Figure 6.14), and the fish becomes either left-eyed or right-eyed. There are regions of the earth where virtually all of the flounders are right-eyed, and regions where they are almost all

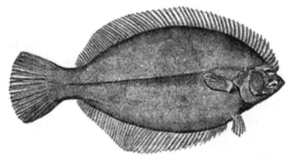

FIGURE 6.14. A flounder (*Pseudopleuronectes americanus*) with two eyes on the same side.

left-eyed. Breeding experiments aimed at determining whether the eyesidedness of flounders was inherited were inconclusive [14]. It could be that the prevailing currents or other characteristics of the particular body of water in which the fish live play an important role in the determination of flounder chirality. This is one example of a bias toward asymmetric structures that always needs to be considered before making broad conclusions concerning animal or plant asymmetry.

HOMO SAPIENS ASYMMETRY

Human beings, like other animals, do not possess perfect bilateral external symmetry. Individuals among us have ear locations that are different on the right or left side, the twirl of our hair can be clockwise or counterclockwise, and we no doubt all have other bumps and other features that are not the same on both sides of the body. You can easily demonstrate asymmetry by taking a mirror and reflecting one half of your face onto the other side, so that your face appears perfectly symmetric. Under this kind of reflection your appearance is markedly different. We would classify most of these asymmetric features as racemic, because they are not reproducible throughout the entire population. There is at least one true asymmetric external feature for men that has been known about since the time of Greek sculptors—namely, the left testicle in most men hangs lower than the right testicle.

Thus far we have completely ignored the asymmetric arrangement of internal organs and the recognized asymmetry of the human brain. Those organs that we have in pairs, such as lungs and kidneys, are symmetrically located, but we have only one heart, one pancreas, one appendix, one large intestine, and so on, and connecting all of these organs together inside the body just couldn't be accomplished while maintaining bilateral symmetry. It also seems to make sense that we would evolve with a genetically programmed organization for placement of the organs, so that the heart and stomach are slightly to the right of center, and the liver is to the left. These plans need to be laid out early in our development, so we get everything where it needs to be. A very rare congenital defect in which placement of the organs is opposite to the normal location, called *situs inversus*, occurs in approximately 0.01% of the population [15].

There are many more examples of asymmetric life in the booklet by Neville [1]. Some of the examples are obvious (fiddler crabs and snails) and some subtle (owl ears, beetle mandibles, etc.). As we walk through our mirror-image forest with an increasingly discerning eye, we see more and more evidence that this world has been reversed. Nothing, perhaps, is more shocking than what we see over the next hill. There is a baseball game being played. Almost all of the players are throwing left-handed, and, oh my, they are running the bases backward (clockwise)!

SUMMARY

A premise of this book is that there exists a hierarchy of symmetry and symmetry constraints in nature. As life evolved and became more complex, more adaptive, more specialized, and perhaps more intelligent, it was necessary to break symmetry constraints. The culmination of this evolution is *Homo sapiens*, where we are dominantly right handed. This is the topic of the next chapter.

SUGGESTIONS FOR FURTHER READING

Neville, A. C., *Animal Asymmetry*, Edward Arnold Publishers, London, 1976, Vol. 67.

Gardner, M., *The New Ambidextrous Universe*, 3rd rev. ed., Dover, Mineola, NY, 2005.

REFERENCES

1. Neville, A. C., *Animal Asymmetry*, Edward Arnold Publishers, London, 1976, Vol. 67.

2. Gardner, M., *The New Ambidextrous Universe*, 3rd rev. ed., Dover, Mineola, NY, 2005.

3. Welch, C. J., Chirality in the natural world: Life through the looking glass, in *Chirality in Natural and Applied Science*, W. J. Lough and I. W. Wainer, eds., Blackwell Publishing, Bodmin, Cornwall, UK, 2002.

4. Palmer, A. R., Antisymmetry, in *Variations*, B. Hallgrimmson and B. K. Hall, eds., Elsevier, 2005, pp. 359–397.

5. Coen, E., *The Art of Genes*, Oxford Univ. Press, New York, 1999.

6. Brown, A. H. and D. K. Chapman, Circumnutation observed without a significant gravitational force in spaceflight, *Science* **225**:230–233 (1984).

7. Edwars, W., A. T. Moles, and P. Franks, The global trend in plant twining direction, *Global Ecol. Biogeogr.* **16**(6):795–800 (2007).

8. Furutani, I., Y. Watanabe, R. Prieto, M. Masukawa, K. Suzuki, K. Naoi, S. Thitamadee, T. Shikanai, and T. Hashimoto, The SPIRAL genes are required for directional control of cell elongation in *Arabidopsis thaliana*, *Development* **127**:4443–4453 (2000).

9. Hashimoto, T., Molecular genetic analysis of left-right handedness in plants, *Phil. Trans. Roy. Soc. Lond. B* **357**:799–808 (2002).

10. Robertson, R., Snail handedness, *Natl. Geogr. Res. Exp.* **9**:120–131 (1993).

11. Gould, S. J., N. D. Young, and B. Kasson, The consequences of being different: Sinistral coiling in Cerion, *Evolution* **39**(6):1364–1379 (1985).

12. Boycott, A. E. and C. Diver, On the inheritance of sinistraility in *Linnaea peregra*, *Proc. Roy. Soc. Lond. B, Biol. Sci.* **95**:207–213 (1923).

13. Schilthuizen, M. and A. Davison, The convoluted evolution of snail chirality, *Naturwissenschaffen* **92**:504–515 (2005).

14. Policnsky, D., The asymmetry of flounders, *Sci. Ame.* 116–122 (May 1982).

15. McKay, D. and G. Blake, Laparoscopic cholecystectomy in situs inversus totalis: A case report, *BMC Surg.* **5**:5 (2005).

16. Chevalier, T., *Girl with a Pearl Earring*, Penguin Books, Middlesex, UK, 1999.

17. Steadman, P., *Vermeer's Camera*, Oxford Univ. Press, Oxford, UK, 2001.

18. Trefethen, L. M., R. W. Bilger, P. T. Fink, R. E. Luxton, and R. I. Tanner, *Nature* **207**:1084–1085 (1965).

19. Shapiro, A. H., *Nature*, **195**:1080–1081 (1962).

LIST OF SIDEBARS AND FIGURES

The Handedness of Homo sapiens

We now return to a discussion of "handedness"! Throughout this book we have used the terms *right-handed* and *left-handed* with an obvious connection to the right and left hands of human beings. We will now focus on the issue of preferred handedness in humans, rather than the use of these terms to simply indicate a particular mirror-image structure. The topic of handedness has been widely studied by psychologists, neuroscientists, biologists, physicians, and many others. This research topic is certainly one in which there are divided opinions, even on some of the most fundamental topics. As we will see, the issue of handedness is closely connected with the asymmetry of our brain, and the diagnostic tools of the twenty-first century have allowed scientists a much clearer view of the physical structure and biochemistry of our brain. In this chapter we shall attempt to present an accurate summary of the voluminous material available on this topic, and highlight the most recent research results.

We should first acknowledge that there have been two very interesting and very detailed books written on this topic. In 1993

Stanley Coren published *The Left-Hander Syndrome: The Causes and Consequences of Left-Handedness* [1], and 11 years later Chris McManus published *Right Hand, Left Hand: The Origins of Asymmetry in Brains, Bodies, Atoms and Cultures* [2]. These two books have been written from the perspective of active researchers in this area, and are full of interesting statistics, anecdotes, and historical perspectives concerning handedness. It is important to note that these authors come to quite different conclusions on several key issues relative to the origin of handedness, including the question of whether being left-handed is programmed into our genes. The reader who wants further information about the issues summarized in this chapter is referred to these books for additional enlightenment.

ARE YOU RIGHT-HANDED OR LEFT-HANDED?

Almost everyone is prepared to answer this question immediately, but there are people who have to think, or maybe qualify their answer. For example, they may write right-handed, but throw a ball left-handed. There have been numerous tests developed to determine whether you are right-handed, or measure how right-handed you are. So answer the following 10 questions, which are based on the Edinburg handedness inventory [3]:

1. Which hand do you use for writing?
2. Which hand do you use for drawing?
3. Which hand do you use for throwing?
4. Which hand do you use when working using scissors?
5. Which hand do you use for brushing your teeth?
6. Which hand do you use for cutting with a knife without using a fork?
7. With which hand do you use a spoon?

8. Which hand do you use to strike a match?

9. Which hand do you use to open the lid on a box?

10. When using a broom, which hand is higher on the broomstick?

If you answered "right" on all 10 questions, you are strongly right-handed, and similarly if you answered "left." Many of you, no doubt, answered mostly with either "right" or "left," and the percentage of your answers that were of your dominant hand is often interpreted as the strength of your tendency to be right- or left-handed.

Some of you rarely or may have never used a matchstick, so how did you answer this question? Tests that have been given to evaluate handedness vary from a few questions to more than 60. Other tests of handedness ask questions about how you fold your arms or clasp your hands, which is your stronger eye or better ear, what foot do you kick with, what eye do you wink with, and so on. Some researchers simply ask the subjects whether they are right- or left-handed, and some other studies make this decision by observing the subject to perform certain tasks, such as throwing a ball and writing. Just about every action that humans take has been analyzed for left–right bias, including the direction in which we turn our heads to kiss. A recent study concluded from observations of kissing in airports, train stations, and parks, that 64.5% of kissing pairs turn their heads to the right to kiss [4].

A very small number of humans are truly ambidextrous. They are equally skilled with either hand, and there are an extremely small number of people who could be classified as lacking the skills that one would associate with being handed at all. The end result of all this testing is that most of us are categorized as being either right-handed or left-handed, some-times qualified with the modifier "strong" or "weak". For a various reasons, some researchers prefer to use the classifications of right-handed and non-right-handed.

THE STATISTICS OF HANDEDNESS

How asymmetric are humans? Of course, this is a bit difficult to answer because of the range of handedness from strong to weak, and because a very few of us are ambidextrous. For this reason, the numbers that you find vary slightly. Just as an example, we will focus on a fairly recent careful study of 628 residents of Mainz, Germany [5]. These subjects were simply asked whether they were right-handed, left-handed, or both, since previous studies had demonstrated that this self-identification correlated extremely well with the Edinburg questions given above. In total, 86.8% of these people said that they were right-handed; 10.5%, left-handed; and 2.7% reported no preferential handedness. This is very similar to many other such observations, although you will find numbers quoted on left-handedness ranging from 5% to 30%. Most references and other sources will say that approximately 90% of humans are right-handed. In general one must be very cautious in the classification of human handedness, and to base any conclusions only on facts (see Sidebar 7.A).

Sidebar 7.A

The British royal family and Google. One of the most important characteristics of the twenty-first century is the almost instant access to information at almost no cost. It is now so easy to explore subjects like handedness on your own, and one can quickly accumulate many interesting facts from thousands of Web pages. A Google® search, for example, of "handedness in the British Royal Family" gets more than 12,000 citations! Of course, whether the information that you get is actually true is not so easy to determine. It doesn't take very many clicks to learn that Queen Elizabeth is left-handed, Prince Charles is left-handed, and Prince Harry, Prince William...the list goes on. In fact, the British Royal family is given as a famous example of the inheritance of left-handedness. It doesn't seem to matter that in a Goggle "image" search there are photographs of the Queen signing documents with her right hand, Prince Charles shooting a basketball with his right hand, and Prince Harry kicking a rugby ball with his right foot. Eventually, you may come across a page of questions and answers from the "Official Royal Website" (www.royal.gov.uk/output/Page2438.asp) dated July 2003 in which you learn that "The Queen, The Duke of Edinburgh (Prince Philip), The Prince of Wales (Charles), Prince Harry, The Duke of York (Prince Andrew), The Earl of Wessex (Prince Edward), and the Princess Royal (Anne) are right-handed. Prince William is left-handed." Should we conclude that the royal family genes are just like ours?

Most researchers would say that there is no obvious geographic dependence on the percentage of right-handed people; however, one should not overlook the cultural dependence of handedness identification. In a very large worldwide study the percentage of left-handers, as measured by the hand that was used for writing, varied significantly between countries such as Mexico (3.5%) and Canada (13.9%) [6]. These results have been interpreted as children being forced to write with their right hand in more "formal" countries such as Mexico, compared to a less formal country like Canada, where this is presumably no longer something that Canadian parents impose on their children [7]. Certainly, one must be cognizant of testing handedness in Islamic countries, where the right hand is considered "clean" and is used for eating and the left hand is used for performing other bodily functions. The negative connotations of being left-handed in some cultures results in some individuals not wanting to self-identify themselves as bring left-handed.

How about footedness? Most people who are right-handed are also right-footed. In the Mainz study, 77.1% of the subjects were right-footed. An analysis of how professional football (soccer) players actually kicked the ball from the World Cup in France in 1998 showed that of the 236 players examined, 77% were right-footed [8]. These statistics are in agreement with those of previous studies. Somewhat surprisingly in this study was the observation that very few players used their feet equally, even though these professional players had been training in and playing this sport for virtually their entire lives. In the Mainz study, 67.8% of the subjects were reported to be right-eyed and 56.6% of the subjects were right-eared. There is certainly a positive correlation between right-handedness and right-earedness, and right-eyedness. So the chances are that if you are right-handed, you are also right-footed, right-eyed, and right-eared. In the Mainz study, women are slightly less likely to be left-handed (9.6%) than men (11.9%). This is also consistent with previous studies. All of the Mainz data excluded individuals with physical or age-related changes that forced one to prefer the use of their right or left side.

There have been numerous studies on the statistics of handedness, and the reader is referred to the list of books and articles at the end of this chapter for further reading. One question that has implications for the evolution of handedness in humans is whether the percentage of left-handed people is increasing or decreasing. These are difficult statistics to obtain, because of the imprecise definitions of handedness, the limited number of years for which data might be available, and the fact that the negative connotations of being left-handed in various societies has led to forcing individuals to use the right hand even though they were left-handed. Furthermore, after someone dies it is impossible, of course, to test them for handedness. Such information can be obtained only from interviewing relatives or friends, or finding photographs where handedness is evident. These indirect methods are naturally much less reliable than a live test.

One good source of data covering the last 130 years is the readily accessible database containing the profiles of American Major League Baseball (MLB) players [9]. This database lists whether the player threw a baseball with his left or right hand, when he was born, and when he died. Using the sole criterion of throwing an object seems to be a reliable indicator of handedness, although it has been argued that the most reliable measure is, in fact, which hand is used for writing [10]. The data for MLB players are given in Figure 7.1, where we plot the percentage of right-hand throwers versus the decade in which they were born.

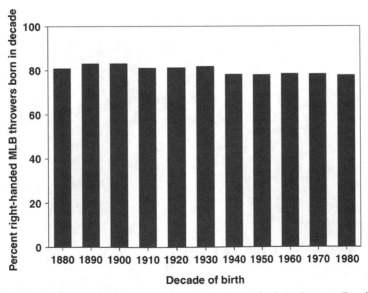

FIGURE 7.1. Plot of the percentage of right-handed Major League Baseball (MLB) throwers versus decade of birth.

There are two obvious observations about these data. First, the percentage of left-handed throwers in baseball is approximately twofold that of the results we introduced earlier from the Mainz study, and this percentage has increased slightly in the last 100 years. Of course, the great thing about baseball is that every statistic like this one is open to lengthy and heated discussion and interpretation by all of us baseball fanatics. Are left-handers simply more athletic? Has the specialized pitching in current baseball provided more opportunities for left-handed relief pitchers? Does the increase in recent years reflect the fact that parents became less inclined to force their children to use the right hand? We leave further interpretation to the interested reader and/or baseball aficionado.

There are more indirect methods to obtain information about handedness from prehistoric times. One recent study analyzed the prehistoric cave drawings of handprints, as shown in Figure 7.2, on the walls of caves in France and Spain [11]. It is presumed that that these handprints were

FIGURE 7.2. A prehistoric cave painting of a right hand. See color insert.

produced by blowing a tube filled with ink onto a hand placed against the wall. The other hand, of course, would be holding the tube. The result from 507 handprints in 26 caves was that 23% of them were of right hands. The researchers in this study compared this result to a group of French college students performing the same task, and the results were almost identical. The French students were less than 10% left-handed as measured by throwing and writing, so not every right-handed student held the tube in their right hand, and the same was assumed for the cave dwellers. These researchers conclude that the percent handedness has not changed significantly in approximately 10,000 years.

There is also evidence for handedness in earlier prehistory. Anterior teeth of Neanderthals have been examined [12]. Scratches on the teeth have been interpreted as being caused by an individual grasping something by his or her teeth while pulling with either the left or right hand. Summary data reveal that of the 20 Neanderthal individuals examined, 18 show evidence of pulling with the right hand and two with the left hand. This result for hominids living approximately 130,000 years ago is remarkably similar (10%) to the percentage of left-handed people living today, although the number of observations is small. Even much earlier observations of stone toolmaking by hominids in Lower Pleistocene sites dating between 1.4 and 1.9 million years ago show a large preference for right-handedness [13]. In this analysis, Toth viewed clockwise rotation of stone flake generation as evidence that the chopping motion used by the hominids to make the stone tools was being performed with the right hand.

THE ORIGIN OF LEFT-HANDEDNESS

Although individual animals such as dogs and cats display a preference for using their right or left paw, there is no evidence for overall footedness in these or any other species. There have been some reports on slight preferences in the way that apes and chimpanzees use certain tools, but these results are not conclusive. It is safe to say that only with *Homo sapiens* do we see a true consistent preference for using our right hand, right foot, and so on. So, why are we so dominantly right-handed, and why aren't we all right-handed? These questions are difficult to answer, but some aspects of these questions are becoming clearer as scientists develop higher-resolution images of the brain, and more capabilities of matching genes to function.

Some controversial conclusions have been made on the basis of various data concerning the minority of people who are left-handed. These conclusions tend to focus on the early development of left-handed individuals, and some even go so far as to speculate that left-handedness is a sign of trauma at birth [14]. The book by Coren bases a significant amount of the discussion on issues associated with a perceived shorter life-span for left-handed people [1]. In the Mainz study mentioned above, the percentage of people who displayed right-side preferences did increase slightly with age. One possible inter-pretation of this age dependence is that older people were, perhaps, more likely to have been forced to use their right hand, so the data might be skewed in that direction. This is another area where it is difficult to obtain reliable data but for which baseball statistics can be used. If one looks at all the MLB players who have died for which we have good birth and death dates and an indication of whether they were a left-handed or a right-handed thrower (a total of 7514 through 2006), one can deter-mine that 18% of these individuals were left-handed. This is just about exactly the percentage of players who were left-handed. The average lifespan of left-handed throwers was 67.7 years, and for right-handed throwers 68.2 years. This seems all too short, but, of course, this includes many players who were born before 1900. Never- theless, one can conclude that there is no significant difference in lifespan for left-handed versus right-handed MLB players.

It should be mentioned that Coren also used baseball statistics to study lifespan differences for left- versus right-handers [15]. He used the book *The Baseball Encyclopedia* from 1975, and defined a strong right-hander as one who both throw right and batted right. This seems to us a curiosity, since many strong right-handers are taught to bat left, because of the advantage of being closer to first base. His database under these definitions totaled 2271. His conclusions concerning left-handed baseball players are very different from those reported above, and he reported significant differences in lifespan for left handers as a function of age. The more recent database that is now available electronically is much larger, and our analysis shows that 17% of those players who died in their nineties or older were left-handed, and approximately 16% of those who lived to be 80 years or more were left-handed. These numbers are so similar to the percentage of left-handers and right-handers in baseball in earlier eras that we fail to see that this database provides support for a shorter lifespan for male athletic left-handers. Coren also used death statistics on 987 people who died in southern California. He obtained informa- tion on handedness by talking to relatives and friends of the deceased, and concluded that on the average right-handers lived 9 years more than did left-handed people. These results have been disputed by a number of researchers for various reasons; probably the most important reason was that he relied on relatives to report on handedness [10].

We now know that handedness is directly associated with asymmetry in the human brain. Fine-motor control for 95% of the human population is controlled by the left hemisphere of the cerebral cortex of the human brain. Most of these people are right-handed, but 70% of left-handed individuals also use the left hemisphere for fine-motor control. The left hemisphere is also the dominant area for language processing (97%), so there has been considerable speculation that somehow language and right-handedness in humans developed at the same time in evolution. Since these are the two major characteristics that set us apart from other animals, this parallel development makes some sense. However, more recent analysis of the speech capabilities of Neanderthals have concluded that their ability to make complicated sounds was very limited [16], and this has led to some speculation that the development of sign language and gestures might be better correlated with the development of brain asymmetry than is spoken language [17].

In the last 25 years, there has been an enormous amount of research performed on brain function, particularly on brain asymmetry. Previously brain researchers had only a limited number of patients with brain deformities or injuries to examine in order to study the functions of different parts of the brain. Now, with modern imaging techniques, it is possible to obtain very high-resolution pictures of the brain in action, so neuroscientists are able to not only correlate specific functions with different regions but also map out more subtle regional physical differences in shape and structure [18]. Even the nonexpert can see that every brain magnetic resonance imaging (MRI) image is asymmetric (see, e.g., Figure 7.3), and it is clear from more expert analysis that the right and left sides of the brain are truly different. Furthermore, this difference is consistent in more than 90% of the population.

THE HUMAN GENETICS OF HANDEDNESS

Certainly right-handedness is an inherited trait. Throughout the regions of the earth and throughout time, humans have been predominantly right-handed. It is not programmed into our genes in the same way that other traits such as having two hands

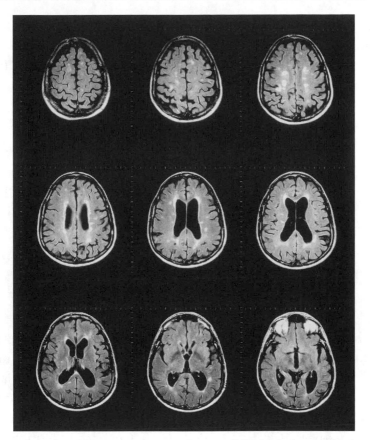

FIGURE 7.3. Magnetic resonance images (MRIs) of a human brain.

or two ears are programmed, but an overall score of 90% for being right-handed is pretty impressive. The bigger question is whether a tendency for left-handedness is inherited. There actually have been many attempts to develop a genetic model for the inheritance of handedness, but to date no one has proposed a model that is universally accepted. Statistics on the handedness of people who are related should be easily obtained since there are more than 5 billion human beings on this planet, but the number of careful studies where relationships were confirmed by blood tests, for instance, have actually been very few. One of the most interesting legends concerning the inheritance of left-handedness has to do with the Kerr family in Scotland, where it was said that left-handedness was much more common than in the normal population. [See Sidebar 7.B].

Sidebar 7.B

Kerr handedness. The following passage memorializing the skill of the left-handed warriors of the Kerr clan comes from a famous poem and drinking song by Walter Laidlaw from 1549 called "The Reprisal" [22]:

> So well the Kerrs their left-hands ply
> The castle gained, the battle won
> The dead and dying round them lie,
> Revenge and slaughter are begun

This poem celebrates the storming and capture of Ferniehurst Castle near St. Andrews in 1548 by Sir John Kerr. This poem and other writings led to the legend that the Kerr clan had an abnormally high number of left-handers. Presumably because they wanted to verify this inheritance of left-handedness, the Royal College of General Practitioners commissioned a study of the people with the surname Kerr (and the anglicized version Carr) in early 1971. The report, which was published in 1974, concluded that, indeed, the Kerr clan had a 29.5% population of left-handed members as opposed to a control group, which had the normal 11% [23]. Shaw and McManus followed up this report with their own study by telephoning 25 people named Kerr and 25 named Carr from the Greater London phone directory [24]. A control group of 48 additional people were also called. The individuals who agreed to participate in the study were asked about themselves and their relatives. and from a total of 1401 individuals, 10.9% (153) were identified as left-handed. The Kerr/Carr participants were actually less likely to be left-handed (9.2%) than the control group (12.7%). These researchers go on to describe in their paper the flawed way in which the Royal Society of General Practitioners study was conducted. GPs were told what the purpose of the study was, and were asked to select some Kerr/Carr members from the patient roles and report on their handedness. Not surprisingly, families that had an above-average number of left-handers were more likely contacted and reported, resulting in a large bias in the final statistical results. This legend is most likely connected to the fact that the Kerr surname is related to the Gaelic word *caerr*, meaning "left" and also "awkward."

Before making any conclusions about the genetics of handedness, or the evaluation of any other model developed to explain the handedness of *Homo sapiens*, it is important to attempt to sort out the statistics about handedness that are generally accepted to be valid, from all the other statistics about handedness that are in dispute. The following four statements are assumed to be valid:

1. The chance that a child will be left-handed doubles when the child's mother is left-handed from a little over 10% to slightly more than 20%, and apparently doesn't depend on the handedness of the child's father.

2. In analysis of data for identical twins, it has been observed that handedness varies among identical twins in approximately the same way as in the general population, independent of the handedness of parents.

3. The percentage of left-handers in the mentally retarded population is 20%, and in the severely mentally retarded, 28% [19].

4. The percentage of members of Mensa (an organization for high-IQ people) who are left-handed is 20%, and left-handed students do proportionately better on the Scholastic Aptitude Test that is given in the United States.

Certainly point 1 above suggests that genetics must be part of the origin of left-handedness, although analysis of confusing data from various studies of twins as shown in point 2 suggests that the genetic component might be quite weak [20]. As mentioned previously, the genetic models that have been developed to explain handedness have not been widely accepted. There is an ongoing search for genes that might be involved in the development of handedness, and there has been a recent report that a gene on human chromosome 2 called LRRTM1 is associated with left-handedness [21]. It will be interesting in the next few years to follow developments in this area.

The overall model for the origin of left-handedness must also explain observations 3 and 4. In this discussion we have tried to avoid the various "sinister" aspects of being left-handed that have affected 10% of the world population. Some aspects of this topic will be mentioned in the next chapter, but we should not completely omit from this discussion the less pleasant aspects of the history of left-handedness. In addition to being forced to use their right hands, left-handers have also been treated as imperfect humans in many ways. Unfortunately, this unfair categorization is still made today in many cultures. A number of researchers in the not too distant past that have said that all left-handers are pathological left-handers, that is to say, that left-handedness was a result of some disease, injury, or other abnormality. Most researchers now believe that the number of left-handed people whose handedness is due to some injury or disease of the left hemisphere is actually quite small.

There is much to be said for the model for handedness proposed by Perelle and Ehrman [10]. They completely dismiss the influence of genetics on handedness, and describe two kinds of right-handedness and three kinds of left-handedness. Their model is based on the premise that brain asymmetry is the major determinant for handedness, and since in 97% of human brains language and fine-motor control are located in the left hemisphere, one would expect the percentage of right-handed people to be 97% if every individual followed the normal pattern. Some individuals who have the mirror-image brain structure might have been "switched," so the two kinds of right-handed people are natural right-handers and "switched" right-handers. In some individuals, certain injuries or diseases may have affected the left hemisphere of the cortex, so that as these people developed, some language and motor skills were restricted to the right hemisphere. These are the pathological left-handers, and maybe this explains the statistics of mental retardation given in point 3 above. In this model there are, of course, a very small number of natural left-handers who have brain structure opposite that of right-handers and have not been switched. The third type of left-hander is what Perelle and Ehrman call the "learned" left-hander. These individuals have the same brain symmetry as do natural right-handers, but at some point in their development began using a toy or some other implement with their left hand. If this handedness was encouraged or just developed on its own, then the brain would have had to locate some language and motor skills in both hemispheres. In this model, some individual left-handers would have developed some additional brain processing skills that might improve their performance on tests and have other advantages that lead to point 4 above. This model is summarized in Figure 7.4.

Although this Perelle–Ehrman model does not involve any genetic component, one could easily add a weak genetic

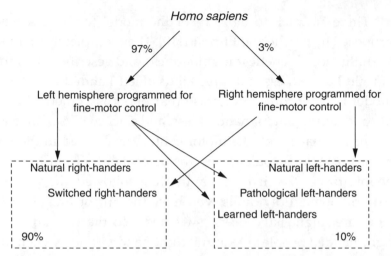

FIGURE 7.4. A schematic diagram of the Perelle–Ehrman model for human handedness.

influence on the probability of the most important issue of brain asymmetry without significantly altering the underlying premise, or major conclusions that could be derived from their model—and, of course, there must be symmetry in this model! If there are learned left-handers from the 97% of left- hemisphere humans, there must be learned right-handers from the 3% of right-hemisphere humans, "pathological" right-handers, and so on. However, since there are such a small percentage of right-hemisphere humans, the inclusion of these cases does not significantly affect the overall statistics of handedness.

SUMMARY

Many volumes have been written and so many millions of dollars have been spent in studying handedness in humans. To attempt to summarize the state of this research in one chapter has been difficult, and no doubt the choice of what to include and what not to include can be debated. We have tried to present some data and conclusions concerning the major issues, the most important of which is whether handedness is inherited. There is no doubt that in the next few years we will see more volumes written and more money spent in studying this topic.

It should be clear that human beings are unique in being so dominantly right-handed, and this uniqueness is directly attributable to an evolutionary specialization of the human brain. It seems that our status as the most intelligent creature on earth is directly connected with our language skills and our being right- or left-handed. Just like the initial choice of L-amino acids, and the fact that most gastropods have dextral shells, and that most twines grow in right-handed helices, there doesn't seem to be any fundamental reason why most of us are right-handed. Is right-handedness just a chance event? Or perhaps some early hominid had a bad fall, and the brain was forced to allocate brain power in a new way, and somehow this asymmetry was passed on to the next generation? Or maybe there was some local chiral environmental influence that resulted in a preference for use of the right hand? Or maybe right-handedness is remotely connected to right-handed helical proteins or right-handed helical DNA? No matter how it happened, the fact that almost all of us are, indeed, right-handed has definitely been reflected in the things we make and use. This is the topic of the next chapter.

SUGGESTIONS FOR FURTHER READING

Coren, S., *The Left-Hander Syndrome: The Causes and Consequences of Left-Handedness*, Vintage Books (Random House), New York, 1993.

McManus, C., *Right Hand, Left Hand: The Origins of Asymmetry in Brains, Bodies, Atoms and Cultures*, Harvard Univ. Press, Cambridge, MA, 2004.

Annett, M., *Handedness and Brain Asymmetry: The Right Shift Theory*, Psychology Press, 2002.

Van Der Velde, C.D., *The Mind: Its Origin and Nature*, Prometheus Books, Amherst, New York, 2004.

REFERENCES

1. Coren, S., *The Left-Hander Syndrome: The Causes and Consequences of Left-Handedness*. Vintage Books (Random House), New York, 1993.

2. McManus, C., *Right Hand, Left Hand: The Origins of Asymmetry in Brains, Bodies, Atoms and Cultures*, Harvard Univ. Press, Cambridge, MA, 2004.

3. Oldfield, R. C., The assessment and analysis of handedness: The Edinburgh Inventory, *Neuropsychologia* 1 (9):97–113 (1971).

4. Güntürkün, O., Human behaviour: Adult persistence of head-turning asymmetry, *Nature* 421:711(2003).

5. Dittmar, M., Functional and pastural lateral preferences in humans: Interrelations and life-span age differences, *Human Biol.* 74(4):569–585 (2002).

6. Perelle, I. B. and L. Ehrman, An international study of human handed-ness, *Behav. Genet.* 24:217–227(1994).

7. Nedland, S. E., I. B. Perelle, V. De Monte, and L. Ehrman, Effects of culture, sex, and age on the distribution of handedness: An evaluation of the sensitivity of three measures of handedness, *Laterality* 9:287–297 (2004).

8. Carey, D. P., G. Smith, D. T. Smith, J. W. Shepherd, J. Skriver, L. Ord, and A. Rutland, Footedness in world soccer: An analysis of France'98, *J. Sports Sci.* 19:855–864(2001).

9. Lahman, S., The Lahman Baseball Database, 2006. [www.baseball1.com]

10. Perelle, I. B. and L. Ehrman, On the other hand, *Behav. Genet.* 35(3):343–350 (2005).

11. Faurie, C. and M. Raymond, Handedness frequency over more than ten thousand years. *Proc. R. Soc. Lond. B (Supp.)*, 271:S43–S45(2004).

12. Fox, C. L. and D. W. Frayer, Non-dietary marks in the anterior dentition of the Kralina Neanderthals, *Int. J. Osteoarcheol.* 7:133–149(1997).

13. Toth, N., Archaeological evidence for preferential right-handedness in the lower and middle Pleistocene, and its possible implications, *J. Human Evol.* 14(6): 607–614(1985).

14. Bakan, P., G. Dibb, and P. Reed, Handedness and birth stress, *Neuropsychologia.* 11:363–366(1973).

15. Coren, E., *The Art of Genes*, Oxford Univ. Press, New York, 1999.

16. Lieberman, P., *The Biology and Evolution of Language*, Harvard Univ. Press, Cambridge, MA, 1984.

17. Corballis, M. C., The gestural origins of language, *Am. Sci.* **87**:138–145, (1999).

18. Sun, T. and C. W. Walsh, Molecular approaches to brain asymmetry and handedness, *Nature Rev. Neurosci.* **7**:655–662(2006).

19. Hicks, R. E. and A. K. Barton, A note on the left-handedness and the severity of mental retardation, *J. Genet. Psychol.*, **127**:323–324(1975).

20. Bishop, D. V., Individual differences in handeness and specific speech and language impairment: Evidence against a genetic link, *Behav. Genet.* **31**:189–198(2001).

21. Francks, C., S. Maegawa, J. Lauren, B. S. Abrahams, A. Velayos-Baeza, and S. E. Medland, LRRTM! on chromosome 2p12 is a maternally suppressed gene that is associated paternally with handedness and schizophrenia, *Molec. Psychiatry* **12**:1129–1139(2007).

22. Bodmer, W. and R. McKie, *The Book of Man: The Human Genome Project and the Quest to Discover Our Genetic Heritage*, Oxford Univ. Press, Oxford, UK, 1997.

23. Research Unit (Royal Society of General Practitioners), *J. Roy. Soc. General Pract.* **35**:437–439(1974).

24. Shaw, D. and I. C. McManus, *Br. J. Psychol.* **84**:545–551(1993).

LIST OF SIDEBARS AND FIGURES

LIVING IN A RIGHT-HANDER'S WORLD

Approximately 90% of all human beings are right-handed. The main topic of this chapter is to illustrate how, in many common situations, this fact has influenced the world in which we live. Those readers of this book who are left-handed certainly know a lot about this subject already from their personal experiences, and this chapter is, perhaps, directed mostly at the right-handed readers to remind or show them how so many aspects of our everyday lives have developed because the majority of us are right-handed. There have been more than a few books written on this subject, and it would be redundant and not very interesting to just provide a list of objects biased for right-handed people. The left-handed or right-handed reader can read one of the available books on this topic, some of which we have listed at the end of this chapter, or, perhaps, stop in at the left-handed kiosk at the Mall of America in Minneapolis or visit one of several websites for the left-hander to purchase pencils, coffee mugs, or other mementos of being left-handed. In Figure 8.1a we show a pencil and a coffee mug designed to enable a left-handed person to read the words while writing

Mirror-Image Asymmetry: An Introduction to the Origin and Consequences of Chirality
by James P. Riehl
Copyright © 2010 John Wiley & Sons, Inc.

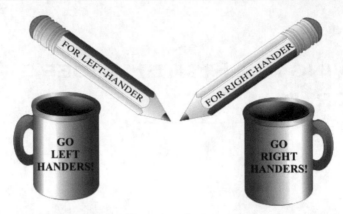

FIGURE 8.1. Pencils and coffee mugs designed for left- and right-handed use.

or drinking with his or her left hand. If these items are used in the right hand, the writing on the coffee mug would be on the "back" side of the mug, and the writing on the pencil would be upside-down and reversed.

In this chapter we will try to highlight other objects or activities in which the right-handed bias is less obvious but clearly present. The influence of being dominantly right-handed affects the way we play sports, on what side of the road we drive our car, the direction in which we write, and the chirality of musical instruments. These will be described along with other topics that illustrate the importance and consequences of asymmetry in our everyday environment.

CLOCKWISE AND COUNTERCLOCKWISE (OR ANTICLOCKWISE) ROTATION

Throughout this book we have used the terms *clockwise* and *counterclockwise* to describe a specific direction of circular motion. Although the circular or rotational motion indicated by the term *clockwise* is a two-dimensional (planar) concept, when you add the observer, it becomes a three-dimensional asymmetric

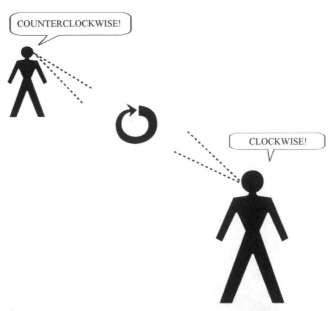

FIGURE 8.2. An observer's view of clockwise versus counterclockwise rotation.

and, therefore, chiral phenomenon. In the United States, the opposite motion is referred to as *counterclockwise*, and in the United Kingdom the common term is *anticlockwise*. Of course, one needs to carefully define from which direction the circular motion is being observed, since a clockwise motion from the front will appear to be a counterclockwise motion from the back as illustrated in Figure 8.2. Adding the numbers to the face of a clock, of course, makes the clock chiral.

The direction that we call *clockwise* originated from the motion of the shadow of an object such as an Egyptian obelisk in the northern hemisphere, and later was the basis for more formalized sundials. Early in the morning the shadow of an object points to the west, and in the late evening the shadow points to the east. At noon the shadow points directly north. This is depicted in Figure 8.3. If the first civilizations that developed measures of time had developed in the southern

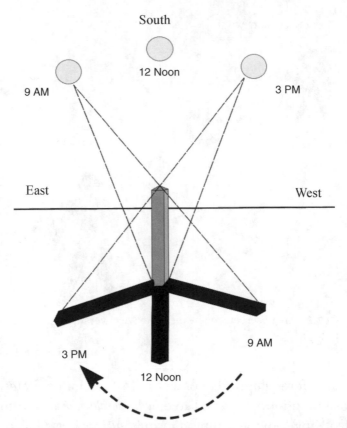

FIGURE 8.3. The clockwise motion of a shadow in the northern hemisphere.

hemisphere, no doubt the direction we call *clockwise* would be reversed. When the first water powered clocks were developed, later followed by pendulum clocks, the face of the clock resembled the sundial (with 12 noon being straight up or directly north), and the direction of the hands followed the same circular motion as was seen for the motion of the shadow in sundials.

Many items in our lives contain objects that need to be turned with our hands. Right-handed human beings find it easier to turn many things clockwise rather than counterclockwise, especially when force must be applied at the same time.

We already know that the definition of a screw that is right-handed helical is one that you need to turn clockwise to have it go *into* an opening. Almost all screws are made in a right-handed helix, and this convention is certainly biased to the advantage of right-handers. This convention applies not only to screws but also to lightbulbs, car keys, and other objects. There are other situations in which the requirement to "turn" or rotate something favors the right-hander. In automobiles, for example, the direction in which the ignition key must be turned is clockwise, and in very old vehicles that needed hand cranking to get started, the direction of cranking, and of course engine rotation as viewed from the front, was clockwise. Old-fashioned (rotary) dial telephones involved rotating the dial in a clockwise direction, and "windup" devices such as clocks or toys are also almost always wound clockwise. In the United States light switches on table and floor lamps invariably require clockwise turning. (European lamps now use mostly symmetric toggle switches on the cords.) Corkscrews are invariably designed for the right-hander, and presenting a right-hander with a left-handed corkscrew to open the wine bottle is always an interesting way to begin a dinner party! Obviously, there is no conspiracy here, it is just that when designing objects such as these one needs to design them for the majority.

HELICAL STAIRWAYS

Helical stairways or staircases are sometimes called *spiral staircases*, although *spiral* in other contexts is a two-dimensional concept. In Figure 8.4 we show the helical staircase attached to the Swenson Science Building on the campus of the University of Minnesota Duluth (UMD). This is a right-handed helical stairway, and students in science are fortunate at UMD because when an instructor wants to show the structure of proteins or DNA, they can just point out the window at the right-handed

FIGURE 8.4. The right-handed helical stairway at the University of Minnesota at Duluth. (Photographed by Brett Groehler.) See color insert.

helix. It is easy to recognize a right-handed helical stairway as one in which, as you climb the staircase, your right hand is on the outside. As you descend a right-hand helical stairway, your right hand is on the inside. Many medieval castles have helical stairs, and very often tour guides at these castles will offer an explanation as to why the stairs were built in the particular helical configuration. It is usually about defending the castle from the marauders who are trying to get up the stairway. The logic is that, since most marauders and defenders are right-handed, their swords are in their right hand. As a defender you would want your right arm to be on the outside of the helical stairs and you want the marauder to have his sword on the inside of the stairs. If this is indeed what you want, then you should make your staircases left-handed helical! This is

FIGURE 8.5. Attacking and defending the Swenson Science building at UMD! (Photographed by Brett Groehler.) See color insert.

illustrated in Figure 8.5 for an attack using the right-handed helical stairway on the Swenson Science Building! In our actual count of castle stairways, it appears that there is little preference for the direction of a helical stairway. Most often in older buildings two staircases are present on opposite sides of the structure, one a left-handed helix and one a right-handed helix. Also, in many large castles the stairways are so wide that handedness wouldn't seem to present a significant advantage or disadvantage to the marauder or the defender.

It is also sometimes said that spiral or helical stairways in ancient churches or monasteries were normally right-handed to allow easier access to the upper levels for right-handers. There doesn't appear to be any reliable statistics on these purported design issues, nor can one find written documentation that the architects of these buildings were really thinking about defense or access when planning the stairways.

In our everyday comings and goings, we don't have many opportunities to use helical stairways, although modern right-helical or left-helical stairways are readily available from various home-supply manufacturers. However, tourists who visit the incredibly beautiful upper chamber of the Sainte-Chapell in Paris with its amazing stained-glass windows will have the experience of climbing a very narrow helical staircase on their way up, and an equally challenging mirror-image helical stairway on their way down. The up staircase on the left as you enter the building is a left-handed helical staircase, and the down staircase on the right is a right-handed helical staircase. The handrails are on the outside, which requires the use of the left hand in both cases, and stepping up or down first with the left foot, since this is where the small wedge-shaped step is the widest. It is not obvious whether this choice of direction is the preferred one for right-handed/right-footed people, but it does seem natural for this right-handed person to step up or down with the left foot when holding a handrail with the left hand.

AROUND THE HOUSE

Although the screwdriver itself is symmetric, we know that tightening a screw is easier for a right-hander. Operation of a power saw is obviously designed for a right-hander, and although scissors might appear symmetric, the application of pressure and twisting is almost always designed for the right hand. In the kitchen, serrated knives, can openers, peelers, corkscrews, and other utensils are usually designed for the right-hander, although left-handed versions are now available. Microwave oven doors open from right to left, with the controls on the right, and stationary electronic mixers are made in an orientation that favors right-handers. Coffee pots and some other appliances generally show preference for the right-handed user. In the office, computer keyboards, especially the location of the numeric keypad, are designed for right-handers, and most left-handers have had to deal with a computer mouse designed for use by the right hand. Camera and camcorders are also designed for right-handed users.

In the garden, starting a chain saw or a gasoline-powered lawnmower requires pulling quickly with the right hand. Chain saws are designed to be operated by right-handers, and can be especially dangerous in the hands of a left-hander. Knitting patterns are written for the right-handed knitter, and if a left-hander chooses to knit left-handed, the knitting pattern will have to be transcribed or read in reverse. Also, yarn is twisted in a right-handed helical fashion, so knitting in reverse might not yield the desired result. This is why many left-handers simply learn how to knit the way right-handers do.

MAKING MUSIC LEFT-HANDED

Although left-handed guitar players such as Paul McCartney and Jimi Hendrix are famous, most musical instruments are designed to be played the way right-handed musicians play them. The guitar and related instruments are unique, because, although left-handed instruments are available, they can be played by left-handed players by simply turning the guitar around, and sometimes restringing the instrument. Left-handed violins are certainly available, but one never sees them used in a large orchestra. It would be a curious sight, indeed, if 10% of the violin, viola, cello, and bass players in an orchestra were bowing in the direction opposite that of the rest of the orchestra.

FIGURE 8.6. A modern French horn.

There is some sentiment among musicians that it might be an advantage to be a left-handed string player, since the fine-motor control required for fingering the instrument is being performed with the dominant hand. So, how could musical instruments develop that seem to give an advantage to the minority left-handed player? To understand this, one should look at the origins of modern instruments. A good example is the horn, commonly called *French horn* in the United States, which is depicted in Figure 8.6. In playing the modern French horn, the valves are manipulated with the left hand, the right hand is placed in the bell of the instrument and does little work, and the position is adjusted to improve pitch and sound quality. Prior to the introduction of valves around 1815, orchestral horns had no valves and players used techniques such as "hand stopping" to change the length of the horn and provide access to notes not on the harmonic series. Since most players were right-handed, they used their right hand for this technique. When valves were added, they were necessarily designed for use by the left hand, which was then being used only to hold the instrument. The horn without valves is called a "natural horn," and although it is very difficult to play, it is still used today when a horn player is interested in reproducing the sound of earlier instruments.

The same kind of history in which right-handed playing led to the need of sophisticated left-handed skills can be seen in the development of string instruments. Modern handheld string instruments originated in the Greek lyre and other instruments in which the strings were of fixed length and plucked with the dominant right hand. Later instruments developed that were played with bows that would be held with the right hand, and when fingerboards and frets developed in such instruments, they would naturally be fingered with the left hand.

THE CHIRALITY OF WRITING

Although two-dimensional objects such as alphabetic letters are not chiral because of the existence of a horizontal symmetry plane, many of them do become chiral when they are written on a piece of paper. So the mirror images of the uppercase English letters AHIMOTUVWXY are virtually identical except for some line thicknesses and minor artistic details associated with a particular choice of fonts, and the lowercase letters are all asymmetric except imovwx. We have chosen to ignore the leading handle on the "m" in this Arial font. These letters all possess a vertical plane of symmetry through their middle. Of course, what really makes writing asymmetric is the stringing of letters together in a specified direction (left to right for those of us from Western civilization) to form words and sentences.

So, how has the fact that we are predominantly right-handed influenced the writing of humans? To address this question we first need to determine what the most common direction for writing is, and then speculate on whether this is a favorable direction for right-handed writers. A very informative Website is maintained by Simon Ager, who, among other things, lists the writing systems that go left to right, right to left, down columns from right to left, down columns from left to right, and other more complicated systems [1]. There are far more left-to-right writing systems than right-to-left, although this doesn't measure how many people actually write in these directions. Chinese and Japanese writing is now either right to left in columns reading down, or left to right in horizontal lines with some variations depending on geography. For example, in Taiwan, Chinese is usually written in vertical columns from right to left, but in mainland China it is written in horizontal lines from left to right.

If you sat down a right-handed person who had not written before, gave her a pen and asked her to try all the different writing directions, which one would she choose? Let's consider only horizontal writing, and decide between right to left and left to right. There do seem to be at least two advantages in writing from left to right. First, you can easily see and read what you have just written. So as you are composing the next word or phrase, you can leave your writing hand on the paper. Also, your writing hand won't smudge the writing just completed, because the part of your hand that rests on the paper is going in the direction away from the recently deposited ink as shown in Figure 8.7. It is speculated that the earliest forms of writing were actually right to left, and exactly why many writing systems switched to left to right is not clear. An interesting case is Egyptian hieroglyphics, in which the direction of writing is variable. (see Sidebar 8.A). Whatever the reason, the current way that Westerners write clearly favors the right-hander. To get his or her hand clear of the writing, the left-handed writer must twist his or her wrist up and around the point where the writing implement is to be placed. This is also shown in Figure 8.7.

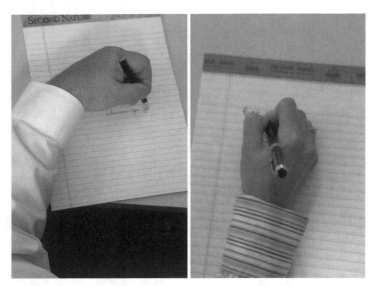

FIGURE 8.7. Left- and right-handed writing. (Photographed by Brett Groehler.)

Sidebar 8.A

Egyptian Hieroglyphics. According to www.omniglot. com, one of the more interesting cases of variability in the direction of a written language is Egyptian hieroglyphics. This language contains more than 2000 characters, some of which are similar to a standard alphabet and correspond to particular sounds, and other characters that could mean either a sound or something associated with the object. The direction of writing was variable, and the text could be written from left to right or right to left in rows or from top to bottom in columns. In order to determine which direction to read a passage of text, one needed to see which direction the animals were facing. So the following phrase, which means "The crocodile sees the cat," could be written as

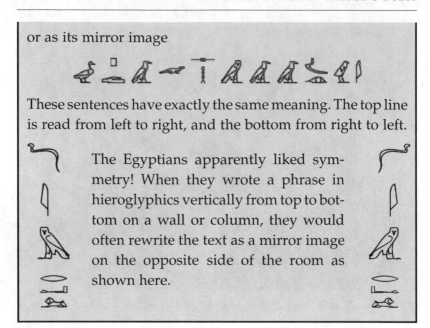

or as its mirror image

These sentences have exactly the same meaning. The top line is read from left to right, and the bottom from right to left.

The Egyptians apparently liked symmetry! When they wrote a phrase in hieroglyphics vertically from top to bottom on a wall or column, they would often rewrite the text as a mirror image on the opposite side of the room as shown here.

SPORTS AND THE LEFT-HANDER

Before beginning our discussion about the advantages or disadvantages of playing sports as a left-hander, we should mention that there are three well-known sports in which playing the game left-handed is forbidden. These are jai alai, field hockey (simply called *hockey* in most parts of the world), and polo. In Figure 8.8 we show representative players in these three sports. Jai alai is a sport that originated in the Basque region of Spain, migrated to Cuba, and is now a popular betting game in Florida. It is played with three walls, and a goat-skin-covered ball about the size of a baseball that is caught and thrown by a wicker-type basket known as a *xistera*. It is sometimes advertised as the "fastest game on earth" since the ball can travel well over 250 km per hour (about 150 miles per hour). In the team version of jai alai there are four players on the court at the same time, and because of the speed of the ball, for safety reasons, it was decided that left-handed playing should not be allowed. In the traditional game of outdoor polo, two teams of four

FIGURE 8.8. Jai alai, field hockey, and polo players.

players each on horseback use a 4-ft-long mallet to strike a 3.25-in.-diameter ball around a playing field of 300×200 yards. (Indoor versions of polo are also played in North America on smaller fields and with fewer players). Scoring involves hitting the ball with the mallet from a mounted position between goalposts 8 yards apart. Because of the combination of mallet swinging and horses galloping, playing left-handed in this sport is also forbidden for safety reasons.

Safety does not seem to be the primary reason why left-handed play is not permitted in field hockey. With a very large field (100 × 60 yards for the outdoor game), there would seem to be plenty of space for the 22 players (11 on each side) to avoid each other's stick. However, there are no left-handed sticks, and hitting the ball with the back rounded side of the stick is not allowed, so you must strike the ball from the right side. Of course, left-handers can play this game; they simply are required to play from the same side as right-handers do (see Sidebar 8.B).

Sidebar 8.B

Playing from the Left Side. The Fédération Internationale de Hockey (FIH) was established in 1924 to formalize the rules of field hockey. Although the rules don't explicitly forbid playing the game left-handed (FIH *Rules of Hockey*, 2004) they have effectively excluded left-handed playing by specifying that

> The stick consists of a straight handle with a curved head. The lower part of the stick's left-hand (playing side) is smooth and flat. The back of the stick (right-hand side or nonplaying side) is smooth and rounded for the entire length of the stick. The ball must be played with the flat side of the stick.
>
> In May 2001, PlanetfieldHockey.com reported the case of Hans Dykstra, who grew up playing field hockey as a left-hander by turning the stick around and hitting the ball in much the same way that we occasionally see professional golfers turning their clubs around and playing left-handed (or right-handed) when their ball is on the "wrong" side of a tree or rock. Hans was apparently an excellent player, and the national coach wanted him on the national team, but the other members of the selection committee decided that he couldn't play internationally because his playing "wasn't normal" and not suitable for international play. Mr. Dykstra played with a Dutch hockey team composed of former international players well into his forties, and was described as the oldest player in the Netherlands playing on a men's first team.

In thinking about the advantages or disadvantages of playing sports left-handed, it is useful to separate the discussion of sports that involve one-on-one interactions between players from the influence of handedness on the overall team sport. The "major" one-on-one confrontations occur in boxing, tennis, and games related to tennis such as squash, badminton, and ping-pong (table tennis). First, we should note that these games are all symmetric. The playing "fields" (or boxing ring) all have planes of symmetry, and if we were observing a match in any of these sports through a mirror, we would not recognize that the game had been reversed. It is usually presumed that left-handed boxers or left-handed tennis players have an advantage because they are fewer of them, and so an opponent is less likely to have experience playing against a left-handed opponent, whereas the left-handed player has a lot of experience playing against right-handers. The records of professional play in these sports, however, don't always bear out this conclusion. For example, in looking at the winners of Wimbledon, 14 of the 120 (12%) men's champions (seven different individuals) have been left-handed, and only 10 of the 113 (9%) women's champions (two different individuals) have been left-handed. (The special athlete Martina Navratilova won nine singles Wimbledon championships!) Similarly, the other major tennis championships do not have a percentage of left-handed champions significantly different from that expected from a population that is 10% left-handed. These percentages and results suggest that it is talent, not left-handedness, that makes good tennis players. Being left-handed might be an advantage at lower levels of competition. Reliable historical data on the handedness of squash and other racket sports are not available.

Boxing is another sport where left-handers supposedly have an advantage, because of the infrequent chances that right-handers have to fight left-handers. The statistics on whether a championship boxer was left-handed are complicated by the fact that boxers sometimes, even during a bout, will switch from a typically left-handed hitting position to a right-handed one. Also, there are so many self-defined "world championships" that it is difficult to decide which are legitimate and how to collect reliable data. In 1994 the *New York Times* did describe Michael Moorer as the "first left-handed heavyweight champion" [2]. If this is true, then certainly left-handers haven't dominated professional boxing. For the reasons given above, it does appear that being left-handed in fencing is a real advantage, since more than 50% of the Fédération Internationale d'Escrime (International Federation of Fencing) world champions are left-handed.

In almost all major team sports, except baseball, which we will discuss below, being left-handed does not seem to be a particular advantage or disadvantage, or to impact how one would play the game. The team sports of ice hockey, American football, soccer, basketball, rugby, cricket, and many others are played on symmetric fields. If one ignores any numbers or letters on the field or on uniforms, then, when viewed through a mirror, the game would appear to be unaffected. Of course, if looking carefully, one might notice an unusual number of left-handed shooters in basketball, left-handed kickers in soccer, and so on, but other than this, the game would be completely recognizable.

Some team sports such as those mentioned above may have tactical or other reasons for locating left-handed or left-footed players on a particular side of the playing field, but these are usually coaching decisions, and are not universally applied. For example, many ice hockey coaches position players who are right-handed shooters on the left wing or left defense to keep their stick in the middle of the ice surface, and try to place their left-handed shooters on the right side. Other coaches vary this coaching strategy. Of course, even in team sports there are one-on-one confrontations, and the usual conclusion is that left-handed players have an advantage for the same reasons given above. For example, in basketball, defending a player who can "go to his left" and shoot with his or her left hand may be difficult for someone who mostly defends right-handers. Similar situations exist in the other team sports.

Some of these handed sports situations have been analyzed from a scientific perspective. For example, researchers have analyzed the number and scoring efficiency of left-handed batting during the World Cup for cricket in 2003 [3]. Overall, 24% of the batters were left-handed, but remarkably almost 50% of the batters of the more successful teams were left-handed. Although left-handed batting does not absolutely define one as left-handed, it is positively correlated with being left-handed. Left-handed batters scored more runs on the average (20 runs per batting appearance vs. 11) and were at-bat for significantly more time than were right-handed batters (25 balls vs. 15 balls). It was concluded by these researchers that, in fact, unfamiliarity with bowling for left-handers by the more inexperienced teams in the world cup accounted for this left-handed bias. Although left-handed bowling was not ana-lyzed in this study, it has been shown that left-handed bowlers tend to be more of the slow-spinning bowlers than are fast bowlers. Why the percentage of successful left-handed batters is so large, and why left-handed bowlers tend to be slow-spinners is not clear.

AMERICAN BASEBALL: THE CHIRAL SPORT

There are very few sports or other human activities that display the evident chirality that is intrinsic in the game of American baseball. We are now again ignoring the issue of numbers and letters on uniforms or on the field. Baseball is the only sport that not only appears completely backward when viewed as a mirror image, but also has specific roles for left-handed players, and specific advantages for left-handed players that are imbed-ded in the nature of the game, and not only in one-on-one interactions between handed players. Perhaps the closest sport to baseball in this aspect would be the game of jai alai mentioned previously. Because this game is played in a three-walled court (front, back, and left) with the right side now commonly used for the viewing/gambling audience, one could easily see that one was watching a mirror-image game, but, of course, there are no left-handed players. A more subtle example of a "chiral sport" would be chess. The serious player would recognize that the playing board on the right in Figure 8.9 is correct but that the

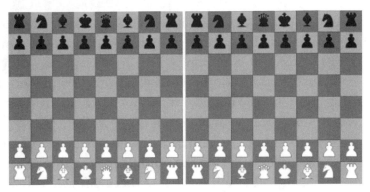

FIGURE 8.9. A chessboard and its mirror image.

mirror-image board on the right side of this figure is not. In chess the convention is that the white square is on the lower right corner ("white on right"), queens occupy the space corresponding to their color (white queen on the white square), and white moves first. These setup rules do make the game chiral, but, of course, there is no advantage or disadvantage in being a left-handed chess player! There are, of course, a huge number of board games and card games in which the order of play is set to be clockwise (or counterclockwise), but these games also provide no specific role or advantage for the left-handed or right-handed player.

Because baseball players run the bases counterclockwise, baseball is a chiral sport, and one that, when viewed as a mirror image, is completely confusing to the baseball observer. The direction of the game also results in left-handed MLB players being essentially excluded from playing the position of third baseman, shortstop, and second baseman. These players must catch the ball and throw it very quickly to first base, which is on their left. A left-handed thrower would have to turn his body after catching the ball to be able to make this throw. Left-handed catchers are also extremely rare in Major League Baseball because of the throwing to bases that is required. There are more right-handed batters than left-handed batters, and the right arm of the catcher is on the side of the plate opposite that of the right-handed batter, making throwing a bit easier. Why is baseball played with a counterclockwise direction of running the bases? Because, of course, humans are mostly right-handed throwers, and although right-handers can and do play every position on the field, left-handed players are more limited. There are no rules prohibiting left-handed players from playing any position, and although there have been a few left-handed players playing traditionally right-handed positions in baseball history, they are very rare. Baseball is surely the only sport that has had to make rules for ambidextrous players! (see Sidebar 8.C.) Certainly when a 3- or 4-year-old child is observed to be a left-handed thrower, and is given his or her first left-handed baseball glove, the child is destined to be a future pitcher, first baseman, or outfielder.

Sidebar 8.C

Switch Pitching. Jerome Holtzman, the Major League Baseball (MLB) historian, reported (4/3/2000, mlb.com) that only one major league baseball player has ever pitched with both hands in a MLB game. The pitcher was Greg Harris of the Montreal Expos, and the date was September 28, 1995 in a game against Cincinnati in the last week of the season. The Expos were 24.5 games out of first place, and the Reds were 9.5 games out in front in the National League Central Division, so the game had no impact on the final season standings. Harris had a special six-finger (two-thumb) glove, which is now in the Baseball Hall of Fame. He pitched to Reggie Sanders as a right-hander (ground out), Hal Morris (walk) and Ed Taubensee (ground out) as a left-hander, and retired Brett Boone on a grounder to the mound as a right-hander to retire the side.

Realizing that batters and pitchers changing sides might develop into a time-consuming standoff, Bobby Brown, President of the American League at that time, sent the following set of rules to his umpires:

1. The pitcher must indicate the hand (right or left) to be used.

2. The pitcher may change arms on the next hitter but must indicate the arm to be used.

3. There will be no warmup pitches between the change of arms.

4. If an arm is injured, the pitcher may change arms and the umpire must be notified of the injury. The injured arm cannot be used again in that game.

Baseball is also the only stadium sport in which an asymmetric playing field is not only tolerated, but actually modern thinking is that the asymmetry adds to the overall interest in the game. Older baseball stadiums such as Fenway Park in Boston with the "Big Green Monster" in right field, or the close right field porch in the old Yankee Stadium in New York City affected not only the style of play but also the selection of players to be used. In the 1960s, many of the old asymmetric stadiums were replaced by modern large symmetric ones such as Busch Stadium in St. Louis or Veterans Stadium in Philadelphia. The construction of Camden Yards Ballpark in Baltimore started a trend back to old-style parks, and all of the new stadiums are being built to exploit and enjoy the chirality of this game.

Even more so than in cricket, there are one-on-one interactions between pitcher and batter in baseball that are influenced by the handedness of the players. This is because of the necessity of running to the right side after hitting the ball as opposed to running straight ahead in cricket. Batting on the right side of

FIGURE 8.10. Batting from the left side in baseball: left-handed batter Chase Utley of the Philadelphia Phillies and right-handed throwing catcher Mike Rabelo of the Florida Marlins.

the plate (as seen by the catcher in Figure 8.10) with your right side toward the pitcher is the normal position for a left-handed batter. Left-handed (or right-side) batters have a clear advantage since they are closer to first base. For this reason many fathers when they are teaching this game to their children have them bat left-handed even if they are right-handed throwers. Because the natural direction of pitching with the left arm and hand is that the pitch will move away from a left-handed batter, making it presumably more difficult to hit, it is a common strategy for the team on the field to try to pitch left-handed pitchers against left-handed batters, and right-handed pitchers against right-handed batters. In general, the statistics bear out this strategy; right-handed batters hit better against left-handed pitchers, and vice versa for left-handed batters.

There are some, perhaps biased American baseball followers who place baseball at the top of the evolutionary hierarchy of sports, just as *Homo sapiens* is at the top of the evolutionary hierarchy of living creatures. Humans have broken the last symmetry constraint and become asymmetric by being 90%

right-handed, and baseball is the only completely asymmetric sport by defining the direction of play to be counterclockwise.

DRIVING ON THE LEFT OR RIGHT SIDE

One of the more interesting aspects of our chiral world on which to speculate is the reason behind the different automobile driving directions. There has been much literature on this subject, and again the interested reader should consult the reference books listed at the end of this chapter. The two interrelated chiral issues associated with driving a vehicle are (1) on which side of the vehicle the driver sits and (2) on which side of the road the person drives. These questions can be answered mostly by realizing that most people are right-handed, but the influence of politics, culture, finances, and other factors have all played a role in getting to where we are today [4]. This is a world in which two-thirds of the population drives their automobiles on the right side of the road. The one-third driving on the left are mostly in countries presently or formerly part of the British Empire. Many of the countries that now drive on the left are island nations (UK, Japan, Australia, and New Zealand), so there currently aren't many situations in which drivers have to change sides of the road at border crossings. This used to be a problem, for example, in Sweden, because in Norway one drove on the right, and in Sweden on the left. At every border crossing, drivers would have to switch sides! After much debate, Sweden changed at 5 AM Sunday morning September 3, 1967 without much trouble (see Sidebar 8.D).

Sidebar 8.D

Driving in Antarctica. Whereas every other country has set uniform rules for which side of the road is used for driving, the continent of Antarctica has no such rule. Obviously, only a very small number of isolated bases are scattered through-out the continent. Each vehicle follows the rules of its home country; however, the New Zealand Scott Base and the McMurdo Station of the United States are only a few kilo-meters apart. So, what does one do when traveling from the Scott Base to McMurdo Station? Peter Kincaid reports in his book *The Rule of the Road* that drivers are simply warned about the driving rules, and in the 26 years prior to publica-tion of his book in 1986, no accidents due to "driving on the wrong side of the road" were reported.

When most "traffic" was solitary walkers or people on horseback, it was customary to travel on the left side of a path or road, since a right-hander's weapon would be on the right side. Also, mounting and dismounting a horse is easier for right-handers from the left side, which would be out of the way of oncoming traffic. When horse-drawn wagons or stage-coaches were used, the driver typically used a whip or lash to drive the team of horses. This would be used in the dominant right hand, so the solitary driver would sit off to the left side or on the rear left horse. This would also be the place to sit to ensure that your wheels were clear of oncoming traffic on your left, so driving on the right side of the road became common. Obvi-ously, this could be confusing, so beginning in the late eigh-teenth century various European countries started to pass laws about what side of the road to drive on, and since most continental European countries legislated driving on the right, drivers became situated on the left side.

In the new United States of America, anti-British sentiment apparently influenced a decision to drive on the right, but in virtually all the images one can find of stagecoaches in the United States in the 1800s, if there is someone else sitting up front with the driver, the driver is sitting on the right side. If the driver was wielding a whip or a lash, then sitting on the right

FIGURE 8.11. Depiction of a US Western stagecoach.

side, even if one was driving on the left, would keep the right hand free for controlling the horses, and also keep the whip away from the body of the front passenger. This is shown in Figure 8.11. When motorized vehicles were invented, of course, whipping or lashing was no longer necessary. What was necessary, however, was changing gears with a typically floor-mounted gearshift. The first gearboxes were no doubt difficult to work, so they were situated in the middle of the front-seat section of the vehicle, for use with the right hand. Drivers then were seated on the left side of the vehicle.

Why do so many countries drive on the left side with the driver situated on the right side? It is difficult to know the answer to this question, but certainly in the United Kingdom a decision in the 1960s not to switch was influenced by the enormous cost of changing road signs, traffic lights, and other such devices. Changing gears in modern cars is, of course, much easier than it used to be, and can certainly be performed with the normally weaker left hand. Although many British subjects would disagree, having to use your left hand for changing gears in an automobile is a little troublesome for a predominantly right-handed population. Of course, many arguments have been made concerning the safety or other aspects of right-side driving. There is at least some thought that having the normally dominant right eye closer to oncoming traffic in Britain results in fewer accidents.

WINDMILLS

Our last example of the influence of right-handers on the world we live in is the chirality of traditional windmills of the type found in The Netherlands and other European countries. The tallest windmill in Holland, "De Valk" (the falcon), located in the city of Leiden, is shown in Figure 8.12. The idea of using windmills for energy was probably brought to Europe by crusaders from the Middle East. Around 1200 AD windmills started appearing in Europe and were providing power to grind grain and drain the lowlands of Holland. By the late 1700s windmills had reached a very high level of standardization and sophistication and, although seen in England, France, Germany, Spain, and Scandinavia, were present mostly in Holland and neighboring areas.

Fortunately, many of the windmills built hundreds of years ago in The Netherlands are still functioning today. Although they are now used primarily as tourist attractions, many still turn, and one can see how they were used for grinding flour or draining ditches. As one can see from Figure 8.12, the blades of a windmill are very similar to the blades of a propeller, so they are certainly chiral, and one can classify them as right-handed or left-handed. As the wind strikes De Valk in Leiden, the blades turn counterclockwise, and following the nomenclature that we

FIGURE 8.12. "De Valk" windmill in Leiden, The Netherlands. See color insert.

introduced earlier, we would say that this is a right-handed propeller. As you travel around Holland on one of the country's many windy days you can see many windmills turning, particularly in some of the more famous tourist areas such as Enkhuizen, where a line of five windmills is operating (see Figure 8.13). If you are observant, you will see that all of these windmills turn in a counterclockwise direction. In fact, all Dutch windmills turn counterclockwise. In fact, all windmills of this type that one might see in Sweden, Ireland, England, and other European countries, or the ones built by Dutch immigrants

FIGURE 8.13. Windmills in Enkhuizen. See color insert.

in the United States also turn counterclockwise. Why do you think that this is so? Why are Dutch windmills homochiral?

Certainly, if you were in the windmill making business, just as in the automobile industry, it would make no sense to manufacture the components of windmills in both left-handed and right-handed chiral forms. So once you started making windmills of one chiral type, you would make them all of the same chirality. But why right-handed? The answer involves knowing a little more about windmills. In Figure 8.14 we show an enlarged view of the blade of a Dutch windmill. There is generally a wooden latticework structure attached to a thick solid beam on which a canvas sail is attached. You can see the beam and the canvas on the right side of this photograph. On windy days, in order to prevent the windmill from turning too quickly, the canvas sail is rolled up and tied to the beam, and on calm days the sail is unfurled and attached to the latticework to help catch more wind. You can see the sails unfurled in Figure 8.13. Blades in these types of windmills always turn in the direction of the beam, namely, counterclockwise. A very important job is to maintain the sails so that the mill is working, but not turning too fast to destroy the inner machinery and

FIGURE 8.14. The blade of a Dutch windmill in Arnhem with the canvas rolled up and secured. See color insert.

gears. To do this the miller climbs up on the latticework of the blade. Imagine yourself assigned to the job of unfurling this canvas as you are looking up as in Figure 8.14! If you are right-handed, you would certainly want to untie the canvas with your right hand and hang on to the latticework with your left hand. For right-handers, the fine-motor skills involved in tying or untying the canvas is something done much more easily with our right hand. We conclude, therefore, that windmills turn counterclockwise, because most people are right-handed!

SUMMARY

Although why most of us are right-handed and not left-handed is not known, our handedness has had a large impact on the nature of the things that we use. We have seen this in chainsaws, automobiles, baseball, and windmills, and if one becomes aware of our chirality, the impact of right-handedness will be evident in many other situations.

SUGGESTIONS FOR FURTHER READING

Kincaid, P., *The Rule of the Road: An International Guide to History and Practice*, Greenwood Press, 1987.

deKay, J. T., *The Left-Handed Book*, M. Evans & Co., 1988.

Lindsay, R., *Left Is Right: The Survival Guide for Living Lefty in a Right-Handed World*, Gilmour House, 1996.

deKay, J. T., *The Natural Superiority of the Left-Hander*, M. Evans & Co., 1979.

Roth, M., *The Left Stuff: How the Left-Handed Have Survived and Thrived in a Right-Handed World*, M. Evans & Co., 2005.

McManus, C., *Right Hand, Left Hand: The Origins of Asymmetry in Brains, Bodies, Atoms and Cultures*, Harvard Univ. Press, Cambridge, MA, 2004.

REFERENCES

1. Ager, S. Omniglot: Writing Systems and Languages of the World, 2007 (available from: www.omniglot.com).
2. Surgery on Moorer's valuable left hand, *New York Times*, May 28, 1994.
3. Brooks, R., L. F. Bussière, M. D. Jennions, and J. Hunt. Sinister strategies succeed at the cricket World Cup, *Proc. Roy. Soc. Lond. B* (Suppl.) **271**: S64–S66 (2004).
4. Kincaid, P., *The Rule of the Road: An International Guide to History and Practice*, Greenwood Press, 1987.

LIST OF SIDEBARS AND FIGURES

Sidebar 8.A. Egyptian hieroglyphics

Sidebar 8.B. Playing from the left side

Sidebar 8.C. Switch pitching

Sidebar 8.D. Driving in Antarctica

THE ASYMMETRIC UNIVERSE

In this book we have seen may examples in which the mirror image of an object is nonsuperimposable on the original object. These examples of chirality span the entire scale from subatomic particles to molecules to humans to planets, and to the universe itself. Our understanding of these mirror-image symmetry relationships began with a study of molecular structures, and the discovery by Louis Pasteur that molecules of tartaric acid existed that were mirror images of each other. This was a critical part of the development of mid-nineteenth-century chemistry, and confirmed new ideas about the three-dimensional structure of molecules. Chemists could make mirror-image (asymmetric) molecules from symmetric starting materials, but it was soon learned that living systems had the remarkable ability to make and use only one of the mirror-image structures. This specific understanding of molecular chirality had to wait almost 100 years to be confirmed with an experimental measurement of a molecular structure through the diffraction of X rays. Nevertheless, the knowledge that asymmetry was so connected to life led Louis Pasteur to state that "I am persuaded that life, as it is known to us, is a direct result of the asymmetry of the universe or of its indirect consequences. The universe is asymmetric."

Once we leave the molecular world of the chemist to explore the atomic and subatomic world of nuclear physics, the living world of biology, or the extraterrestrial world of astronomy, we do not find and therefore cannot compare exact mirror-image structures, although we still view many aspects of the world as approximately symmetric. It is often much easier to describe objects and processes in terms of deviations from symmetric, rather than attempt to understand the full asymmetry. An obvious example of this is ourselves. Human beings may look symmetric, but we are certainly chiral asymmetric structures. Our left sides are not identical in appearance to our right sides, we are almost all right-handed, our internal organs are distributed in an organized but asymmetric manner, our proteins and DNA are twisted predominantly as right-handed

FIGURE 9.1. The Taj Mahal in Agra, India.

helices, the sugar molecules that make up our DNA are all D-enantiomers, the amino acids that are bound together to make the necessary proteins of life are all L-enantiomers, and the atoms that make up everything are held together by chiral nuclear forces. Even the symmetric molecules that we draw as a tetrahedron (methane), hexagon (benzene), or "buckyball" (buckminsterfullerene) are simply idealized structures, since the molecules are in constant motion with vibrations, rotations, and translations. At any point in time the structures are surely asymmetric and chiral.

Although we know that the universe is asymmetric, we seem more comfortable with objects that are symmetric. We see beauty in the Taj Mahal (Figure 9.1), the Washington Monument, the Eifel Tower, and the St. Louis Arch, and we are

FIGURE 9.2. The Walt Disney Concert Hall in Los Angeles, California.

certainly challenged by the asymmetry of a building designed by renowned architect Frank Gehry such as the Walt Disney Concert Hall in Los Angeles (Figure 9.2). However, it is in fact the challenges of asymmetry that can make life more interesting and the discovery of the causes of asymmetry that are often the most rewarding for those of us trying to understand the universe.

Young scientists-to-be should see the universe as something of wonder, as well as something waiting to be explored and waiting to be understood. There is so much to learn about the asymmetry of our universe, but to paraphrase Albert Einstein; the asymmetric universe is so amazingly comprehensible.

LIST OF FIGURES

Figure 9.1. The Taj Mahal in Agra, India.

Figure 9.2. The Walt Disney Concert Hall in Los Angeles, California.

UNDERSTANDING CHEMICAL STRUCTURE DRAWINGS

For more than 150 years chemists have been using drawings to communicate information about chemical structures. The system that all modern chemists use shows which atoms are connected to which other atoms, and depending on the context, often tries to show important aspects of the three-dimensional structure of molecules. There is far too much chemical knowledge built into a chemical structural drawing to describe here, and the intent of this appendix is to simply help the reader unfamiliar with chemical structures understand the most important features of these drawings.

The element of primary interest to many chemists and biochemists is carbon (element symbol C). Carbon always makes four bonds, so we could draw the molecule methane, which has four hydrogen atoms (H) attached to carbon as follows, where the solid lines are used to indicate bonds:

$$H-\underset{\underset{\displaystyle H}{|}}{\overset{\overset{\displaystyle H}{|}}{C}}-H$$

Mirror-Image Asymmetry: An Introduction to the Origin and Consequences of Chirality by James P. Riehl
Copyright © 2010 John Wiley & Sons, Inc.

This two-dimensional structure gives no 3D information. The structure of methane (CH_4) is actually tetrahedral, so if a chemist needs to communicate this information, the structure would be drawn as follows, where the bonds that are in the plane of the paper are drawn as straight lines, those that come out of the paper are drawn as solid wedges, and those that are directed below the paper are drawn as dashed wedges:

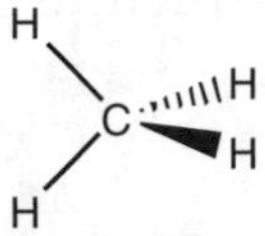

Current convention for dashed wedges places the narrow end of the wedge in the plane of the paper, and the wide end behind the plane of the paper, but many chemists will draw the dashed wedges with the narrow end pointing below the paper to indicate an increasing distance from the observer.

Carbon forms many compounds where multiple bonding exists between carbon and an attached atom. These are called *double* and *triple bonds* and are present in the molecules carbon dioxide (CO_2) and acetylene C_2H_2 as drawn below:

Note that carbon has four bonds in each structure. These molecules are both linear, and the drawing is done to illustrate this fact. Another compound in which carbon has multiple bonds is benzene, which is a planar molecule and is drawn below in two different ways:

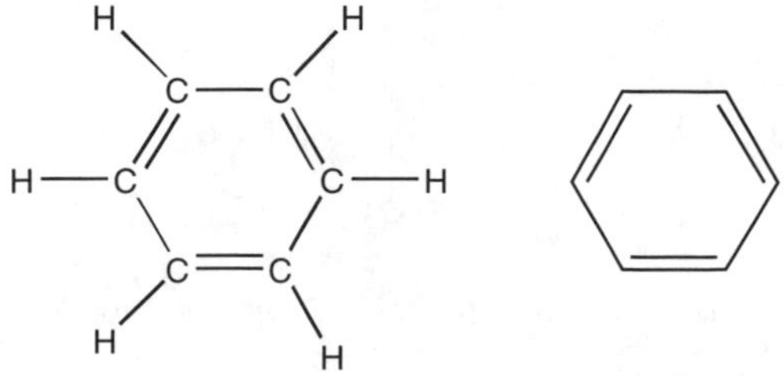

As you can see from the drawing on the right, the symbols for carbon and hydrogen are no longer present. For large structures containing many carbon and hydrogen atoms, it becomes too confusing and unnecessary to show all the atomic symbols, and a convention has arisen that the end of any solid or dashed line or wedge, or the intersection of lines that doesn't contain an elemental symbol, is assumed to be carbon. Furthermore, for every carbon atom in the structure it is assumed that there are enough H atoms attached so that the total number of bonds is four. The result of these drawing conventions can be seen in the benzene structures given above, and we draw three more examples below:

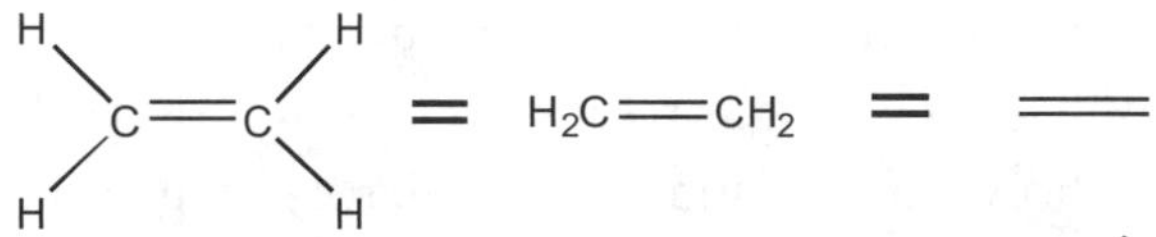

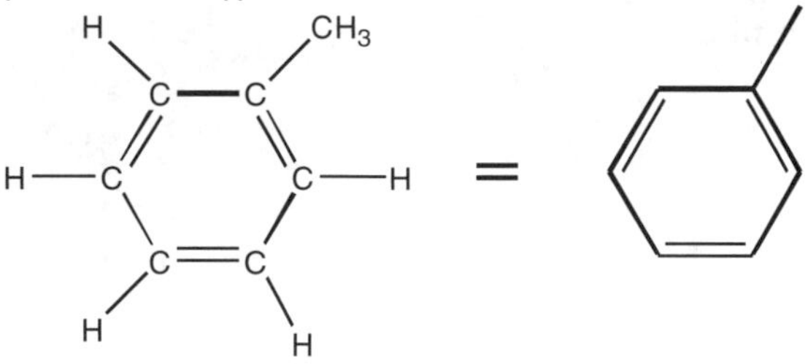

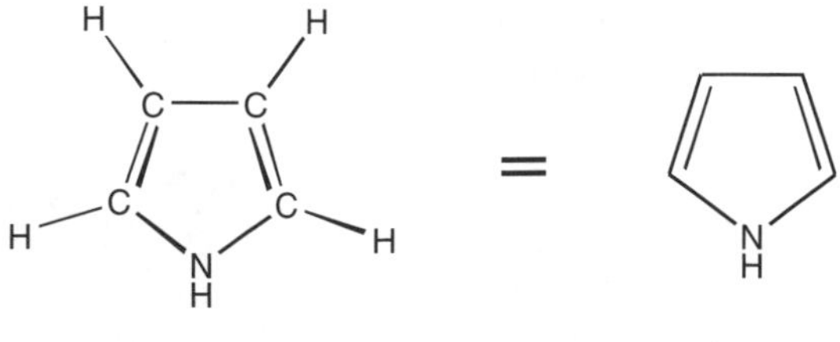

One other piece of information to help one see some of the geometry of the molecules drawn in this book is to show how chemists draw nonplanar rings. The most common example is the six-carbon ring seen in cyclohexane, which has the formula C_6H_{12}:

The structure on the far left, of course, does not imply any three-dimensional geometry, and certainly a chemist would know that this is not a planar molecule. The middle and right structures do try to show the structure of this molecule in three dimensions. This particular conformation is known as the "chair" form of cyclohexane. Perhaps the "chair" reference can be seen in the structure on the right? Hopefully, readers of this book who are not familiar with chemical drawings have learned enough in these few pages to appreciate the complexity of the molecules shown, and visualize the mirror-image relationships of chiral centers.

Biography

Dr. James P. Riehl grew up in Philadelphia, Pennsylvania, and received a B.S.
degree in Chemistry from Villanova
University in 1990, and a Ph.D. in Physical Chemistry from Purdue University in
1975. After a postdoctoral fellowship at
the University of Virginia, he joined the
Chemistry faculty at the University of

Missouri—St. Louis, where he progressed through the academic ranks, being promoted to Professor of Chemistry in 1991. In
1993 he accepted the position as Chairman of the Department of
Chemistry at Michigan Technological University, and in 2000
he was appointed to his current position as Dean of the
Swenson College of Science and Engineering and Professor of
Chemistry at the University of Minnesota Duluth. He has
also held visiting faculty positions at Kings College, London,
and the University of Leiden, The Netherlands.

Dr. Riehl is the author or coauthor of more than 100 research
publications. His research interests are in the theoretical and
experimental application of spectroscopic methods to probe
the structure and dynamics of chiral molecules. In 1992 he

received the St. Louis Award of the American Chemical Society for his "outstanding contributions to the chemistry profession." He has more recently been recognized for his combination of research and administrative accomplishments by being awarded a University of Minnesota McKnight Presidential Leadership Endowed Chair. In 2005 he was awarded the Gold Medal of the University of Wroclaw in Poland for his longstanding efforts at developing research and education partnerships with this university.

INDEX